AF893519

CODES, DESIGNS AND GEOMETRY

edited by

Vladimir Tonchev
Michigan Technological University

A Special Issue of
DESIGNS, CODES AND CRYPTOGRAPHY
An International Journal
Volume 9, No. 1 (1996)

KLUWER ACADEMIC PUBLISHERS
Boston / Dordrecht / London

DESIGN, CODES AND CRYPTOGRAPHY

An International Journal

Volume 9, No. 1, August 1996

Special Issue: Designs, Codes and Geometries

Guest Editor: Vladimir Tonchev

Distributors for North America:
Kluwer Academic Publishers
101 Philip Drive
Assinippi Park
Norwell, Massachusetts 02061 USA

Distributors for all other countries:
Kluwer Academic Publishers Group
Distribution Centre
Post Office Box 322
3300 AH Dordrecht, THE NETHERLANDS

Library of Congress Cataloging-in-Publication Data

A C.I.P. Catalogue record for this book is available
from the Library of Congress.

Copyright © 1996 by Kluwer Academic Publishers

All rights reserved. No part of this publication may be reproduced, stored in a retrieval system or transmitted in any form or by any means, mechanical, photo-copying, recording, or otherwise, without the prior written permission of the publisher, Kluwer Academic Publishers, 101 Philip Drive, Assinippi Park, Norwell, Massachusetts 02061

Printed on acid-free paper.

Printed in the United States of America

Design, Codes and Cryptography, 9, 5–6 (1996)
© 1996 Kluwer Academic Publishers, Boston. Manufactured in The Netherlands.

Preface

This volume contains papers based on talks presented at the Second Upper Michigan Combinatorics Workshop on Designs, Codes and Geometries held on the campus of Michigan Technological University, Houghton, Michigan from August 17th to August 20th, 1994. The conference was attended by 55 participants including 18 supported graduate students, and featured eight invited addresses (*) and twenty contributed talks:

Ed Assmus* (Lehigh University), *Some aspects of the coding theory of designs.*
Elaine Bell (Auburn University), *Decomposing the complete graph into cycles.*
Jurgen Bierbrauer (Michigan Technological University), *Perpendicular arrays.*
Iliya Bluskov (University of Victoria, Canada), *Designs with maximally different blocks.*
Jeffrey Bonn (Southern Illinois University at Carbondale), *Linear greedy codes over* $GF(q)$.
Laurel Carpenter (Clemson University), *Some results on codes of oval designs.*
Anie Chaderjian (Scripps College), *New classes of residue supplementary difference systems.*
Rodolfo Chi (National University of Mexico), *On Caley's construction and certain block designs.*
Charles Colbourn* (University of Waterloo, Canada), *Making a difference (matrix).*
Gary Ebert (University of Delaware), *A group theoretic characterization of Buekenhout-Metz unitals.*
Anant Godbole (Michigan Technological University), *Covering arrays: upper bounds and Poisson approximations.*
Dawit Haile (Southern Illinois University at Carbondale), *Minimal edge-critical graphs with a given small girth.*
Yuri Ionin (Central Michigan University), *Maximal intersecting families with constant intersection size.*
Jennifer Key* (Clemson University), *Codes and finite geometries.*
Hiroshi Kimura (Ehime University, Japan), *Hadamard matrices and dihedral groups.*
Clement Lam* (Concordia University, Canada), *BDX - A computer program to explore block designs.*
Spyros Magliveras* (University of Nebraska-Lincoln), *Block transitive resolutions of designs and Room rectangles.*
Laura Monroe (University of Illinois at Chicago), *Self-orthogonal greedy codes.*
Iosif Pinelis (Michigan Technological University), *On the minimal number of even submatrices of 0-1 matrices.*
Vera Pless* (University of Illinois at Chicago), *Coding constructions.*
Mohan Shrikhande (Central Michigan University), *Multilinear polynomials and designs.*
Mark Stamp (National Security Agency), *Politically correct sequences.*
Deborah Street* (University of New South Wales, Australia), *Constructions of change-over designs.*

Jef Thas* (University of Gent, Belgium), *Arcs, hyperovals, and flocks.*
Tran van Trung (University of Essen, Germany), *Ordered Steiner systems.*
Wal Wallis (Southern Illinois University at Carbondale), *Totally balanced arrays and triple systems.*
Judy Walker (University of Illinois at Urbana-Champaign), *A new approach to the main conjecture on algebraic-geometric MDS codes.*
Qing Xiang (Ohio State University), *Reversible Abelian Hadamard difference sets and projective three-weight codes.*

The conference organizers, Alphonse Baartmans, Donald Kreher, and Vladimir Tonchev, gratefully acknowledge the support from the National Science Foundation, the Institute of Combinatorics and its Applications, and the Michigan Technological University.

Design, Codes and Cryptography, 9, 7–27 (1996)
© 1996 Kluwer Academic Publishers, Boston. Manufactured in The Netherlands.

Designs and Codes: An update

E. F. ASSMUS, JR.
Department of Mathematics, Lehigh University, Bethlehem PA 18015, U.S.A.

J. D. KEY*
Department of Mathematical Sciences, Clemson University, Clemson SC 29634, U.S.A.

Communicated by V. Tonchev
Received February 2, 1995; Accepted August 15, 1995

Abstract. Since the publication in 1992 of "Designs and their Codes" significant progress has been made in the general area of codes coming from designs. This article reviews this progress and presents some of the results — including confirmation of certain conjectures made and answers to some of the questions raised in the book.

Keywords: Designs, codes, finite geometries

1. Introduction

"Designs and their Codes" [3] described the association of linear error-correcting codes with designs and collected together many of the significant known results. Recall that the linear p-ary code associated with a design $\mathcal{D}$ is equivalent to the row space over the finite field $\mathbf{F}_p$ (where p is a prime) of an incidence matrix for $\mathcal{D}$ with rows indexed by the blocks of $\mathcal{D}$ and columns indexed by the points. Various classes of designs were discussed in the book and the corresponding results concerning the codes were described and established. A second printing — with corrections but substantially unchanged — appeared in 1993; here we will review some of the classes of designs discussed in the book — updating the results — and also examine further cases not covered there. We arrange this survey into the following ten sections:

1. Introduction
2. Basic definitions and terminology
3. Projective and affine planes
4. Oval designs
5. Hadamard designs
6. Unitals

* Support of NSF grant GER-9450080 acknowledged

7. Steiner triple systems
8. Finite geometries and rigidity theorems
9. Dual structures and the "point" code of a design
10. Strongly regular graphs and their p-ranks

Notice that the survey is restricted to results related to codes associated with designs; thus there are some notable omissions concerning progress in design theory, in particular with respect to difference sets, where there have been important developments. (See [18] and [72] for a survey of recent results concerning difference sets.) We have, moreover, included nothing concerning the theory of Reed-Muller and generalized Reed-Muller codes, the subject of Chapter 5 of [3], since we have already prepared an updated version of that chapter which is available as a report [5] and will appear in *The Handbook of Coding Theory* — which is to be published by Elsevier.

2. Basic definitions and terminology

Our notation is quite standard and we refer the reader to the Glossary of [3]. We recall, however, some of the definitions we need.

An incidence structure $\mathcal{D}$ consisting of a block set $\mathcal{B}$ on a point set $\mathcal{P}$ of size v is a t-(v,k,λ) design — or t-design for short — if every block is incident with precisely k points and any set of t distinct points are together incident with precisely λ blocks. When $|\mathcal{B}| = |\mathcal{P}|$, t is at most 2 and, when $t = 2$, we refer to the *symmetric* design as a (v, k, λ) design. It follows (see [3, Chapter 1]) that a t-design is also an s-design for any $s < t$; we denote the number of blocks incident with s points by λ_s. The order of a t-design, where $t \geq 2$, is $n = \lambda_1 - \lambda_2$. Because of the statistical origins of the theory, λ_1 is frequently denoted by r and λ_0 by b. We take the *incidence matrix* of an incidence structure to be the $|\mathcal{B}| \times |\mathcal{P}|$ matrix of 0s and 1s where the entry indexed by (B, x) is 1 if and only if the block B is incident with the point x.

For any field F, $F^{\mathcal{P}}$ is the vector space of functions from $\mathcal{P}$ to F with basis given by the characteristic functions of the singleton subsets of $\mathcal{P}$. If $\mathcal{D}$ is an incidence structure, the **code** $C_F(\mathcal{D})$ of $\mathcal{D}$ over F is the subspace of $F^{\mathcal{P}}$ spanned by the characteristic functions, v^B with B in $\mathcal{B}$, of the blocks of $\mathcal{D}$. Viewed concretely this code is simply the row space of an incidence matrix of the incidence structure. If $F = \mathbf{F}_p$ we write $C_p(\mathcal{D})$, or sometimes merely $C(\mathcal{D})$ if p is understood, and the dimension of $C_p(\mathcal{D})$ is referred to as the **p-rank** of $\mathcal{D}$. It is a well-known result, stated and proved in [3, Theorem 2.4.1], that the prime p must divide the order of the 2-design for the p-ary code of the design to be of any use or interest in any characterization. We wish to emphasize, however, that recent results indicate that in some cases, in particular in the case of Steiner triple systems, the column space of the incidence matrix — i.e. the code of the dual structure (see below) — is of interest for *any* prime p.

In considering the code C associated with a structure, it is necessary also to consider the orthogonal code $C^{\perp}$, where, in this update, the orthogonal is always taken with respect to the standard inner product, namely $(u, w) = \sum_{x \in \mathcal{P}} u(x)w(x)$ for $u, w \in F^{\mathcal{P}}$. Thus

$$C^{\perp} = \{u \mid u \in F^{\mathcal{P}} \text{ and } (u, w) = 0 \text{ for all } w \in C\}.$$

Recall that for a design $\mathcal{D}$ and a field F, the following codes, closely related to $C_F(\mathcal{D})$, sometimes play an important role

$$C_F(\mathcal{D}) \cap C_F(\mathcal{D})^{\perp} = \mathrm{Hull}_F(\mathcal{D}) = H_F(\mathcal{D}),$$

$$C_F(\mathcal{D}) + C_F(\mathcal{D})^{\perp} = \mathrm{Hull}_F(\mathcal{D})^{\perp} = B_F(\mathcal{D}),$$

and

$$\langle v^B - v^C | B, C \in \mathcal{B}\rangle = \langle v^B - v^{B_0} | B \in \mathcal{B}\rangle = E_F(\mathcal{D}),$$

where B_0 is some fixed vector in $\mathcal{B}$. If F is $\mathbf{F}_p$, then H_p, B_p, and E_p have the obvious meaning.

Also of interest is the code of the dual structure $\mathcal{D}^t$ where, if $\mathcal{D}$ has block set $\mathcal{B}$ and point set $\mathcal{P}$, then $\mathcal{D}^t$ has block set $\mathcal{P}$ and point set $\mathcal{B}$ with a "block" x incident with a "point" B precisely when B was incident with x in $\mathcal{D}$. Concretely it is the column space of an incidence matrix for $\mathcal{D}$ which is the same as the row space of an incidence matrix for $\mathcal{D}^t$ since these matrices are transposes of one another. Even when $\mathcal{D}$ is a symmetric design the codes $C_F(\mathcal{D})$ and $C_F(\mathcal{D}^t)$ may not be isomorphic although they do, of course, have the same block length and dimension.

Recall that the weight of a vector is the number of non-zero coordinates. Clearly the code of a design will have minimum weight at most the block size k and the code of the dual structure will have minimum weight at most $\lambda_1 = r$.

3. Projective and affine planes

Finite projective planes of order n are symmetric designs with parameters $(n^2+n+1, n+1, 1)$ and finite affine planes are designs with parameters 2-$(n^2, n, 1)$, where again n is the order of the plane.

The basic general results concerning codes of planes are developed in [3, Chapter 6], and we state here the main results.

THEOREM 1 *Let* Π *be a projective plane of order* n *and let* p *be a prime dividing* n. *Set* $H = \mathrm{Hull}_p(\Pi)$ *and* $B = H^{\perp}$. *If* k *is the dimension of* $C_p(\Pi)$, *then* B *is an* $[n^2+n+1, n^2+n+2-k, n+1]$ *code and the minimum-weight vectors of* B *are precisely the scalar multiples of the incidence vectors of the lines of* Π. *Moreover,* $\mathrm{Hull}_p(\Pi) = E_p(\Pi)$.

Thus the plane can be retrieved from the minimum-weight vectors of $C_p(\Pi)$ and hence different planes must have different codes. There has been little progress in the main outstanding conjectures in this area, in particular the Hamada-Sachar conjecture (see [3, Conjecture 6.9.1]):

Conjecture 1 *Every projective plane of order p^t, p a prime, has p-rank at least $\binom{p+1}{2}^t + 1$ with equality if and only if it is desarguesian.*

The Hamada-Sachar conjecture remains the most important conjecture; its truth would imply that the only planes of prime order are the desarguesian planes since, in the prime case, the p-rank of any projective plane of prime order p is $\binom{p+1}{2} + 1$.

Concerning the orthogonal code, the general result is

THEOREM 2 *If Π is a projective plane of order n and p is a prime dividing n, then the minimum weight of $C_p(\Pi)^\perp$ is at least $n+2$. If the minimum weight is $n+2$ then n is even, $p = 2$ and the minimum-weight vectors are all incidence vectors of ovals in Π.*

There are projective planes of even order — namely certain non-translation planes of order 16 — containing no ovals[1] (see [55]). Of course, the classical desarguesian planes always have ovals, namely those coming from conics, but for orders greater than 8 there are others and there has been progress both in producing and classifying these (see, for example, [53]). Moreover we know that the ovals generate the dual of the plane in this case: see Theorem 9 below.

The general situation for affine planes is outlined in the following theorem:

THEOREM 3 *If π is an affine plane of order n and p is a prime dividing n, then the minimum weight of $B = \mathrm{Hull}_p(\pi)^\perp$ is n and all minimum-weight vectors are constant* [2]. *Further,*

$$\mathrm{Hull}_p(\pi) = \langle v^L - v^M \mid L \text{ and } M \text{ parallel lines of } \pi \rangle.$$

We are left with the question concerning the nature of the minimum-weight vectors of $C_p(\pi)$, where π is an arbitrary affine plane. We know that the scalar multiples of the incidence vectors of the lines of π are among the minimum-weight vectors and that they are the only minimum-weight vectors when π is of prime order; moreover, when π is desarguesian these vectors are, again, the only minimum-weight vectors (see Theorem 5 below). Gordon Royle [59] has found a way to isolate the vectors of weight 9 in the ternary codes of each of the six non-desarguesian affine planes of order 9 and has determined that only the scalar multiples of the incidence vectors of lines occur. Knowledge of the Baer subplanes in each of these planes confirms this, since, by Corollary 6.4.2 of [3, page 211], any weight-9 vector in the ternary code of the plane must be the incidence vector of a line or a Baer subplane. It might thus be tempting to conjecture that the minimum-weight vectors of the code of any affine plane of order the square of a prime must be scalar multiples of the incidence vectors of the lines but we are not aware of any further evidence for this. The general case is still undecided. It is perhaps worth mentioning here that none of the weight distributions of the ternary codes of the seven affine planes of order nine has been computed. Here are the known results in the desarguesian case:

THEOREM 4 *Let p be any prime, $q = p^t$, and $\Pi = PG_2(\mathbf{F}_q)$. Then $C_p(\Pi)$ is a generalized Reed-Muller code and has dimension $\binom{p+1}{2}^t + 1$. The minimum-weight vectors of $C_p(\Pi)$ are the scalar multiples of the incidence vectors of the lines. Further,* $\mathrm{Hull}_p(\Pi)$ *has* *minimum weight $2q$ with the minimum-weight vectors the scalar multiples of the differences of the incidence vectors of distinct lines of Π. The minimum weight $d^{\perp}$ of $C_p(\Pi)^{\perp}$ satisfies*

$$q+p \leq d^{\perp} \leq 2q,$$

with equality at the lower bound if $p=2$.

For affine planes we have

THEOREM 5 *If $\pi = AG_2(\mathbf{F}_q)$, then $C_p(\pi)$ is a generalized Reed-Muller code and has dimension $\binom{p+1}{2}^t$. The minimum weight of $C_p(\pi)$ is q, and the minimum-weight vectors of $C_p(\pi)$ are the scalar multiples of the incidence vectors of the lines of π. Further,* $\mathrm{Hull}_p(\pi)$ *has minimum weight $2q$, with the minimum-weight vectors the scalar multiples of the differences of the incidence vectors of distinct parallel lines in π. The minimum weight $d^{\perp}$ of $C_p(\pi)^{\perp}$ satisfies*

$$q+p \leq d^{\perp} \leq 2q,$$

with equality at the lower bound when $p=2$.

Question 1 *If $\mathcal{D}$ is a finite projective or affine plane of order n, and p is a prime dividing n, what is the minimum weight of $Hull_p(\mathcal{D})$ and what is the nature of the minimum-weight vectors?*

We believe that the minimum weight is $2n$, but we have no idea how one might prove this. A projective plane with the property that the only minimum-weight vectors are the vectors $v^L - v^M$ where L and M are lines, was called "tame" in [3]. Desarguesian planes are tame and it is possible that only the desarguesian planes are tame, but we have no evidence for this.

The enumeration of translation planes has been pushed to order 49 by Rudi Mathon and Gordon Royle, [44]. All the 7-ranks have been computed and, curiously, all are odd. None meets the bound, namely 1135, given in Theorem 6.8.1 of [3]. We are indebted to Gordon Royle for the table of ranks, Table 1.

There have been some interesting developments in the study of the codes of desarguesian planes and we now turn to these. In a paper primarily devoted to nets, Moorhouse [47] established a basis for the affine desarguesian plane of prime order p, and hence also for the projective desarguesian planes of prime order. Dougherty [26] extended Moorhouse's results concerning nets. Notice that in the prime case the hull of the plane is equal to the orthogonal code for both the projective and affine cases.

THEOREM 6 *Let π denote the desarguesian affine plane $AG_2(\mathbf{F}_p)$ of prime order p. A basis for the code $C_p(\pi)$ can be found by taking the incidence vectors of the following lines: all the p lines from any one parallel class; any $p-1$ lines from any other parallel class; and so on, until a single line is chosen from one of the final two parallel classes, and no lines are chosen from the remaining class. This gives*

$$p+(p-1)+(p-2)+\cdots+1 = \frac{1}{2}p(p+1) = \binom{p+1}{2}$$

Table 1. The 7-ranks of the translation planes of order 49

7-rank	Number		7-rank	Number
785	1		919	1
855	1		921	7
897	1		925	4
899	2		927	4
901	1		929	5
905	1		931	12
907	5		933	4
911	2		935	18
913	1		937	25
915	5		939	179
917	4		941	1064

lines, whose incidence vectors form a basis for $C_p(\pi)$.

Thus we have

COROLLARY 1 *With notation as in the theorem, a basis for $C_p(\pi)^\perp$ that consists of minimum-weight vectors can be found by taking the difference of pairs of incidence vectors of lines chosen as follows: choose any one line from each of the parallel classes chosen as in the theorem; for each class take the pair consisting of the chosen line and each other line chosen for the basis. This gives*

$$(p-1)+(p-2)+\cdots+1+0=\frac{1}{2}p(p-1)=\binom{p+1}{2}-p$$

vectors that form a basis.

Since the code of the projective plane can be found by simply adding the line at infinity and its points, the following is immediate:

COROLLARY 2 *Let Π denote the desarguesian projective plane $PG_2(\mathbf{F}_p)$ of prime order p. A basis for the code $C_p(\Pi)$ can be found by taking the incidence vectors of the following lines: choose any one line L; then take all the other p lines through any one point; any $p-1$ lines from a second point on L; and so on, until a single line is chosen from one of the final two points, and no lines are chosen through the last point. This gives*

$$1+p+(p-1)+(p-2)+\cdots+1=\frac{1}{2}p(p+1)+1=\binom{p+1}{2}+1$$

lines, whose incidence vectors form a basis for $C_p(\Pi)$.

Further, a basis for $C_p(\Pi)^\perp$ that consists of minimum-weight vectors can be found by taking all the differences of the incidence vectors of each line chosen as above, starting from the second line, with the first, L. This gives

$$p+(p-1)+(p-2)+\cdots+1=\frac{1}{2}p(p+1)=\binom{p+1}{2}$$

vectors that form a basis.

Recently Blokhuis and Moorhouse [7] found a basis of minimum-weight vectors for $PG_2(\mathbf{F}_p)$ (and hence for the affine case as well), involving any conic in the plane.

THEOREM 7 *Let Π denote the desarguesian projective plane $PG_2(\mathbf{F}_p)$ of prime order p, and let C denote a conic in Π. Then a basis for the code $C_p(\Pi)$ can be found by taking the incidence vectors of all nonsecants to C, i.e. all tangents and exterior lines.*

A basis for $C_p(\Pi)^\perp$ can be found by taking the complements of the incidence vectors of the secants.

The basis for $C^\perp$ given in the theorem is not in terms of minimum-weight vectors, but that can be found again quite simply by taking all the differences $v^L - v^{L_0}$, where L_0 is some chosen tangent or exterior line. Similarly it is easy to get a basis in the affine case using a conic.

This paper has more to say about generators in the general case:

THEOREM 8 *Let Π denote the desarguesian projective plane $PG_2(\mathbf{F}_q)$ of order q, a power of a prime p. If C is a conic in Π, then $C_p(\Pi)$ is spanned by the nonsecants (i.e. tangents and exterior lines) of C, and $H_p(\Pi)$ is spanned by the complements of the secants, and also by the complements of the nonsecants.*

In the case of p odd, all arcs of $q+1$ points are conics, by a theorem of Segre [60]; however, in the case $p = 2$, there are non-regular ovals whenever $q \geq 16$. Computations with Cayley [9] by Carpenter [16] using non-regular ovals in cases up to $q = 64$ lead us to believe that Theorem 8 is true for any $(q+1)$-arc for $q = 2^m$, all m. Note in this case Pott [58] has shown that the incidence vectors of ovals span the orthogonal code, so that a basis of minimum-weight vectors could be found in this case also:

THEOREM 9 *Let Π be a finite projective plane of even order with an abelian group G acting regularly. If Π has ovals then the incidence vectors of the orbit of a oval under G span the orthogonal code $C_2(\Pi)^\perp$.*

The result of Pott answers the query posed in [3, Section 6.6, page 219]. In fact, the only planes known with regular abelian groups *are* the desarguesian planes.

4. Oval designs

There is a well-known class of Steiner systems (i.e. designs with $\lambda = 1$) which, following Wertheimer [70], we will call *oval designs*. Bose and Shrikhande first described these designs in [8]. Recall that by an oval in a projective plane of even order n we mean a set of $n+2$ points that meets each line of the plane in 0 or 2 points.

Definition. Let Π be a projective plane of order $n = 2k$ and let $\mathcal{O}$ be an oval of Π. The **oval design** $W(\Pi, \mathcal{O})$ is the incidence structure with points the lines of Π exterior to $\mathcal{O}$ and blocks the points of Π not on the oval $\mathcal{O}$; incidence is given by the incidence in Π.

That this is a Steiner system is easy to show:

PROPOSITION 1 *The incidence structure $W(\Pi, \mathcal{O})$ is a 2-$(2k^2 - k, k, 1)$ design of order $n = 2k$.*

In [3, Chapter 8] the following result concerning the 2-rank of an oval design is proved:

THEOREM 10 *Let Π be a projective plane of even order n and let $\mathcal{O}$ be an oval of Π. Let $W(\Pi, \mathcal{O})$ be the associated oval design. Then*

$$\text{rank}_2(W(\Pi, \mathcal{O})) \leq \text{rank}_2(\Pi) - (n+1).$$

If $\Pi = PG_2(2^m)$, we have that $\text{rank}_2(W(\Pi, \mathcal{O})) \leq 3^m - 2^m$.

In the desarguesian case $\text{rank}_2(\Pi) = 3^m + 1$ which yields the last inequality above.

If Π is desarguesian with $n = 2^m$, then regular ovals (consisting of the $n+1$ points of a conic plus its nucleus) always exist, and the adopted notation for the oval design from a regular oval is then $W(2^m)$, following [13], due to the association with Witt: see [13, Section 2.6] for an outline of the history of these designs.

In answer to a conjecture in Mackenzie [41] (see also [3, Section 8.4, page 304]) Carpenter [16] has shown the following, using the result of Blokhuis and Moorhouse:

THEOREM 11 *If Π is desarguesian with $n = 2^m$, and $\mathcal{O}$ is any regular oval then*

$$\text{rank}_2(W(\Pi, \mathcal{O})) = \text{rank}(W(2^m)) = 3^m - 2^m.$$

Conjecture 2 *If Π is desarguesian with $n = 2^m$, and $\mathcal{O}$ is any oval then*

$$\text{rank}_2(W(\Pi, \mathcal{O})) = 3^m - 2^m.$$

Carpenter [16] has verified this conjecture for all the ovals in the desarguesian planes of orders less than 64 — and for some ovals in the desarguesian plane of order 64. There still does not seem to be a coding-theoretic way of distinguishing a regular oval from a sporadic oval — as there is in the case of unitals, where the hermitian unitals do distinguish themselves by being contained in the code of the containing plane: see [3, Theorem 6.7.1, page 226] — and this lends at least some meager support to the conjecture.

5. Hadamard designs

The following question was posed in [3, Section 7.11, page 284] and until very recently no counter-example had been found:

Question 2 *Does the binary code of a 3-$(2^m, 2^{m-1}, 2^{m-2} - 1)$ design always contain a copy of the first-order Reed-Muller code, $\mathcal{R}(1, m)$?*

Many classes of Hadamard designs yield codes that do contain copies of the first-order Reed-Muller code and recently Carpenter [16, 17] has found this to be true for a class of designs related to regular ovals. We will discuss these classes first and then exhibit the example yielding a negative answer in the general case.

Any oval $\mathcal{O}$ in a projective plane of even order n can be used to define a Hadamard 2-design $\mathcal{E}$ in the following way: for the points of $\mathcal{E}$ take the $n^2 - 1$ exterior points to $\mathcal{O}$; for each point x of $\mathcal{E}$ define a block B_x to be the set of points

$$B_x = \{y \mid y \text{ is an exterior point and } xy \text{ is a secant to } \mathcal{O}\} \cup \{x\}.$$

This construction gives a 2-$(n^2 - 1, \frac{1}{2}n^2 - 1, \frac{1}{4}n^2 - 1)$ Hadamard design that extends uniquely to a 3-$(n^2, \frac{1}{2}n^2, \frac{1}{4}n^2 - 1)$ design. (A more general way to construct the Hadamard designs is described in [3, Section 7.12]; alternatively this design can be described using the block graph of the Steiner 2-design. See also Maschietti [42] for further descriptions.)

L. L. Carpenter [16, 17] has now confirmed that the designs obtained from regular ovals in the desarguesian planes of even order contain copies of the first-order Reed-Muller code:

THEOREM 12 *Let* Π *be the desarguesian projective plane of order* 2^m, *where* $m \geq 1$, *and let* $\mathcal{O}$ *be a regular oval. If* $\mathcal{T}$ *is the Hadamard design constructed as described above from* $W(\Pi, \mathcal{O}) = W(2^m)$, *then* $C_2(\mathcal{T})$ *contains a copy of the first-order Reed-Muller code* $\mathcal{R}(1, 2m)$. *Furthermore, sets of* $4(2^m - 1)$ *blocks of* $\mathcal{T}$ *can be found that generate* $\mathcal{R}(1, 2m)$ *and are thus common to* $\mathcal{T}$ *and a copy of the affine-geometry design of points and hyperplanes in* $AG_{2m}(\mathbf{F}_2)$.

For the first non-trivial case, i.e. $m = 2$, $C_2(\mathcal{T})$ *is* $\mathcal{R}(1, 4)$, but this never occurs again for regular ovals: see [42].

A further conjecture, raised in Mackenzie [41] and quoted in [3, Section 7.12, page 292], concerns the rank of these designs:

Conjecture 3 *If* $\mathcal{T}$ *is the Hadamard design constructed by the method described above from* $W(2^m)$ *then* $\mathrm{rank}_2(\mathcal{T}) = 2^{m-1}m + 1$.

Tom Norwood [49] recently used a classification by Jackson [34] of Hadamard designs with $SL_2(\mathbf{F}_q)$ acting transitively to prove this conjecture.

Returning now to Question 2, the following construction has recently been shown, through computational results of Gordon Royle, to yield a counter-example: let $\mathcal{D}$ be the hermitian unital on 28 points, a 2-$(28, 4, 1)$ design. The block graph of this design yields a Hadamard 2-design $\mathcal{E}$ with parameters 2-$(63, 32, 16)$. The complementary design $\overline{\mathcal{E}}$ extends to a Hadamard 3-(64,32,15) design, $\mathcal{T}$, with 2-rank 15. We claim that $C_2(\mathcal{T}) \not\supset \mathcal{R}(1, 6)$. The binary code $C_2(\mathcal{E})$ is an even-weight code of dimension 14, and was found to have 7497 vectors of weight 32. The search for the even-weight subcode of the punctured code in $C_2(\mathcal{E})$ was done by first collecting the 7497 vectors of weight 32 and observing that under the action of the automorphism group $P\Gamma U_3(\mathbf{F}_9)$ they fall into two orbits of length 3024, and one each of length 1008, 378 and 63 (the blocks). Searching then for six vectors that generate the even-weight subcode of $\mathcal{R}(1, 6)^*$ was computationally possible, and the search failed.

The weaker question

Question 3 *Is the binary code of a* 3-$(2^m, 2^{m-1}, 2^{m-2} - 1)$ *design always inside a copy of* $\mathcal{R}(m - 2, m)$?

has not been negated and may still have a positive answer.

A further question that was posed in [3, Question 7.10.1, page 282] is also still unanswered:

Question 4 *Does the binary code of a*

$$(2^{2m}, 2^{2m-1} - 2^{m-1}, 2^{2m-2} - 2^{m-1})$$

design contain the all-one vector?

We mention that all the designs of these parameters that can be constructed from any oval design in a plane of even order in the manner described in [3, Section 7.12] do contain the all-one vector.

Designs with parameters $(2^{2m}, 2^{2m-1} - 2^{m-1}, 2^{2m-2} - 2^{m-1})$ are intimately connected with bent functions (see [3, Section 7.9]) and a theorem of Dillon and Schatz, [3, Theorem 7.10.4], which characterizes those designs with 2-rank $2m + 2$ and binary code containing the all-one vector. Recently Deirdre Smeltzer [63] has shown that if a difference set D in an elementary abelian 2-group is given by a bent function on the $2m$ variables $x_1, \ldots, x_m, y_1, \ldots, y_m$ of the form

$$\sum_{i=1}^{m} x_i y_i + h(x_1, \ldots, x_m),$$

where h is any cubic, and if C is the code over $\mathbf{F}_2$ of the development, $\mathcal{D}$, of D (see [3, Theorem 7.10.3]) then $E_2(\mathcal{D})$ is *isomorphic* to the first-order Reed-Muller code $R = \mathcal{R}(1, 2m)$. It follows, of course, that the dimension of C is $2m + 2$ and that the all-one vector is in the code. Therefore the Dillon-Schatz Theorem applies, as pointed out by Dillon [24]. Thus $\mathcal{D}$ has the symmetric difference property. Dillon also showed that $\mathcal{D}$ is, in fact, isomorphic to the development of the difference set given by the quadratic bent function $q = \sum_{i=1}^{m} z_i w_i$ — which is the same as the design given by the minimum-weight vectors of $R \cup (q + R)$.

Another result concerning codes of Hadamard designs is due to T. S. Michael who shows in [46] that the incidence matrix of a $(4n - 1, 2n - 1, n - 1)$ design from any skew Hadamard matrix has maximal p-rank, namely $2n$, when p divides n. The proof involves only elementary linear algebra. The result shows, in particular, that a Hadamard matrix cannot have a skew form unless its p-rank is $2n$ for all odd primes p dividing n (see [3, Theorems 7.4.1 and 7.4.2]).

6. Unitals

A unital, or unitary design, is a Steiner 2-design with parameters 2-$(m^3 + 1, m + 1, 1)$. Apart from one with $m = 6$ constructed by Mathon [43] and independently by Bagchi and Bagchi [6], all other known unitals have m a prime power. There are two known that have a regular cycle acting on points, one of which is the design with $m = 6$; the other has $m = 4$ and is one of the cyclic designs in [6]. Shobe [61] has examined the codes associated with these cyclic designs. The order of a unitary design is $m^2 - 1$, so the prime p over which the code is defined must have p dividing $m - 1$ or $m + 1$. In fact we know of no example where p dividing $m - 1$ and not $m + 1$ (and thus $p \neq 2$) gives a code that is not the full space. For the two classes of 2-transitive unitals this is known to be the case: see Mortimer [48]; hence in these "classical cases" one takes p dividing $m + 1$. For the cyclic unitals found in [6], the 2-$(217, 7, 1)$ design has 7-rank 214; the 2-$(65, 5, 1)$ design has 5-rank 63.

In [3] it was mentioned that more concerning the p-ary codes of the hermitian unitals on $q^3 + 1$ points, where p divides $q + 1$, was expected to be announced. A complete general theory has not been established, but Geck [29] has published some results on the Brauer characters of the unitary group, and Hiss [32] has communicated some partial results to us. Using Magma [15] we have computed the p-ranks of the hermitian unitals for all q such that $q \leq 13$, and we have found that all these codes, for p dividing $q + 1$, have dimension $b/q = (q^2 - q + 1)q$, where b is the number of blocks of the unital. This is the formula suggested originally by Andriamanalimanana [1] based on computations up to and including $q = 5$. The further computations up to $q = 13$ now lead us to formally state this as a conjecture, which is also borne out by the results mentioned in [29] and [32]:

Conjecture 4 *Let $\mathcal{H}$ be the hermitian unital on $q^3 + 1$ points. If p is any prime dividing $q + 1$, then the p-rank of $\mathcal{H}$ is $(q^2 - q + 1)q$.*

Magma also leads us to the following conjecture:

Conjecture 5 *Let $\mathcal{H}$ be the hermitian unital on $q^3 + 1$ points, where q is odd. Let $C = C_2(\mathcal{H})$. Then*

$$\mathrm{Hull}_2(\mathcal{H}) = \begin{cases} C^{\perp} & \text{if } q \equiv 3 \pmod 4 \\ <\jmath> & \text{if } q \equiv 1 \pmod 4 \end{cases}$$

where $\jmath$ is the all-one vector.

Magma has verified this for $q \leq 13$. Similarly, this conjecture does not contradict indications coming from the study of the modular characters of unitary groups, as indicated in [32].

We note here also another property — possibly related to the observations above — that a search with Magma has brought out. Whenever q is odd there is the so-called Hölz design with parameters 2-$(q^3+1, q+1, q+2)$ whose blocks consist of all the blocks of the unital together with the intersections of all Baer subplanes of the ambient plane of order q^2 that meet the unital in $q+1$ points; the computations concern the binary codes of these associated Hölz designs (discovered by Hölz [33] and discussed in [3, Section 8.3]). Looking at the binary codes of the Hölz designs with Magma we found, for odd q between 3 and 13, that the code is the same as that of the unital when $q \equiv 3 \pmod 4$, and larger (probably $\langle \jmath \rangle^{\perp}$) when $q \equiv 1 \pmod 4$.

The unitals with $m = 3$, namely the 2-$(28, 4, 1)$ designs, have received a great deal of attention, partly because of the existence of two such designs having a doubly-transitive automorphism group when m is any odd power of 3 (see [3, pp. 301-302] and [2]), but also because they are small enough to examine in full. Brouwer [11] had already produced many such designs and had investigated when they could be embedded in projective planes of order nine. Gordon Royle [54] has since determined that precisely 17 of these unitals can be so embedded, one of these in two distinct planes. (There are precisely four projective planes of order nine.) As far as we know there has been no progress in characterizing the Ree unital on 28 points as the one of smallest 2-rank; its 2-rank is 19.[3]

7. Steiner triple systems

A Steiner triple system is a 2-$(v,3,1)$ design. Doyen, Hubaut and Vandensavel [27] proved the following result giving a lower bound for the p-rank of any Steiner triple system, and showing that only $p=2$ or $p=3$ need be considered, when looking at the code.

THEOREM 13 *Let $\mathcal{D}$ be a 2-$(v,3,1)$ design and let p be a prime. Then*

(1) $\text{rank}_p(\mathcal{D}) = v$ *if* $p \geq 5$;

(2)

$$\text{rank}_2(\mathcal{D}) = v - d_P - 1 \geq v - \log_2(v+1),$$

with equality if and only if $\mathcal{D}$ is the design of points and lines of the projective geometry $PG_d(\mathbf{F}_2)$, where $v = 2^{d+1}-1$;

(3)

$$\text{rank}_3(\mathcal{D}) = v - d_A - 1 \geq v - \log_3(v) - 1,$$

with equality if and only if $\mathcal{D}$ is the design of points and lines of the affine geometry $AG_d(\mathbf{F}_3)$, where $v = 3^d$.

(Here d_P and d_A are the projective and affine dimensions, respectively, of $\mathcal{D}$: see Teirlinck [65].)

This theorem is in itself in the nature of a rigidity theorem, in that it characterizes the designs associated with finite geometries through the p-ranks of their incidence matrices. Since perfect linear single-error-correcting codes over any field are unique up to a monomial transformation, the binary code of the design of points and lines of a projective space over $\mathbf{F}_2$ characterizes this design. But more is true as was shown by Key and Sullivan [40] for the binary case, and by Key [37] for the ternary case. Here are the relevant results.

THEOREM 14 *Let $\mathcal{D}$ be a 2-$(v,3,1)$ design with $v \geq 7$. Let d satisfy $2^d - 1 \leq v < 2^{d+1}-1$. Then $C_2(\mathcal{D})$ contains a subcode C that can be shortened (by removing $v-(2^d-1)$ coordinate places where all the codewords of C are zero) to a code that is isomorphic to the binary Hamming code $\mathcal{H}_d$. In particular, if $v = 2^d - 1$ then $C_2(\mathcal{D}) \supseteq \mathcal{H}_d$.*

Equivalently, $C_2(\mathcal{D})$ contains a set of weight-3 vectors whose supports form the blocks of the design of points and lines of $PG_{d-1}(\mathbf{F}_2)$.

This result extends to the quadruple systems:

THEOREM 15 *If $\mathcal{E}$ is a Steiner quadruple system, i.e. a 3-$(v,4,1)$ design, with $2^d \leq v < 2^{d+1}$, then the binary code $C_2(\mathcal{E})$ contains a subcode that can be shortened to the Reed-Muller code, $\mathcal{R}(d-2,d)$. In particular, if $v = 2^d \geq 8$ then*

$$C_2(\mathcal{E})^\perp \subseteq \mathcal{R}(1,d) \subseteq \mathcal{R}(d-2,d) \subseteq C_2(\mathcal{E}).$$

In the ternary case we need only consider the triple systems, since, by a theorem of Dehon [19], the ternary code of a quadruple system is the entire ambient space.

THEOREM 16 *Let $\mathcal{D}$ be a 2-$(v,3,1)$ design with $v \geq 9$. Suppose that d is such that $3^d \leq v < 3^{d+1}$. Then $C_3(\mathcal{D})$ contains a subcode C that can be shortened (by removing $v-3^d$ coordinate places where all the codewords of C are zero) to a code that is isomorphic to the generalized Reed-Muller code $\mathcal{R}_{\mathbf{F}_3}(2(d-1),d)$. In particular, if $v = 3^d$ and $d \geq 2$, then we always have*

$$C_3(\mathcal{D})^\perp \subseteq \mathcal{R}_{\mathbf{F}_3}(1,d) \subseteq \mathcal{R}_{\mathbf{F}_3}(2(d-1),d) \subseteq C_3(\mathcal{D}).$$

Equivalently, $C_3(\mathcal{D})$ contains a set of weight-3 vectors whose supports form the blocks of the design of points and lines of $AG_d(\mathbf{F}_3)$.

There is some interesting work in progress concerning Steiner triple systems and *non-linear* binary perfect codes. The problems being discussed go all the way back to Steiner's original paper [64]; it was clear early on (see [4]) that any perfect binary code containing the zero vector yielded an extendable Steiner triple system and that Steiner's questions were intimately related to perfect codes. Moreover, it is still widely believed that every Steiner triple system is extendable and this has been verified for the 80 systems on 15 points [23]. By the theorem of Doyen et al. the Steiner triple systems of smallest possible rank, being classical systems, are extendable. A bit more can be said with respect to this question: using the codes of the systems and the results mentioned above, the following proposition is proved in [38, 39]:

PROPOSITION 2 *Let $\mathcal{D}$ be a Steiner triple system on $v = 2^d - 1$ (or 3^d) points and suppose that the 2-rank of $\mathcal{D}$ is $2^d - d$ (respectively, the 3-rank is $3^d - d$). Then $\mathcal{D}$ can be extended to a Steiner quadruple system. In the binary case the extension can be defined in such a way that its binary code is the extended binary code of $\mathcal{D}$.*

Thus, not only are the classical systems extendable but also those on a classical number of points with rank just above the classical rank.

A related question concerning binary perfect codes asks whether every Steiner triple system on $2^m - 1$ points can be seen as the set of weight-3 vectors of a perfect[4] binary code containing the zero vector [56]. Progress in showing this for the 80 systems on 15 points is being made. The number of inequivalent Steiner triple systems is known rather precisely in an asymptotic sense [71] and Phelps [57] has produced an asmpytotic lower bound on the number of inequivalent binary perfect codes (see also Etzion and Vardy [28]). There are more than enough perfect codes to justify the question. Of course, a given triple system may appear as the set of weight-3 vectors in many inequivalent perfect binary codes and a given perfect binary code may yield many (at most 2^{n-m} where $n = 2^m - 1$) inequivalent Steiner triple systems. Indeed, the number of perfect binary codes grows so quickly with increasing block length that it appears certain that any triple system that does appear is likely to complete to a perfect code in an enormous number of ways. One can see the phenomenon even for small lengths and mere extensions of triple systems; for example, there are two triple systems on 13 points but four quadruple systems on 14 points and each of the triple systems extends in more than one way to a quadruple system (see [45]). For the systems on 15 points this phenomenon is even more striking (see [57]).

8. Finite geometries and rigidity theorems

Hamada's conjecture from [31] still stands in a weakened form, *viz.*:

Conjecture 6 *If a design has the same parameters as a design of points and fixed-dimensional subspaces or flats of a projective or affine space over a field of characteristic p, then the p-rank of the design will be at least that of the corresponding geometric design.*

This conjectures that the designs coming from the geometry have the smallest p-rank, and this has not been shown to be false, although the examples of Tonchev [66] and of Delsarte and Goethals [30] show that it is possible for designs with the parameters of a geometric design to have the *same* p-rank as the geometric design.

The conjecture is most interesting in the hyperplane case and here it is still open in its stronger form:

Conjecture 7 *A symmetric design with parameters*

$$\left(\frac{q^{n+1}-1}{q-1}, \frac{q^n-1}{q-1}, \frac{q^{n-1}-1}{q-1}\right)$$

where $q = p^s$, p a prime, has p-rank at least

$$\binom{n+p-1}{n}^s + 1$$

with equality if and only if it is the design of points and hyperplanes of $PG_n(\mathbf{F}_q)$.

The conjecture has only been proven for $q = 2$ (see Theorem 17 below).

The symmetric design of points and hyperplanes of a projective geometry has a doubly-transitive automorphism group acting. All (v, k, λ) designs with a doubly-transitive group acting and $2k < v$ are known [36]; besides the projective designs there is another infinite class with parameters $(2^{2m}, 2^{2m-1} - 2^{m-1}, 2^{2m-2} - 2^{m-1})$ and two sporadic examples: the unique $(11, 5, 2)$ design and the so-called Higman design with parameters $(176, 50, 14)$. The designs of the second infinite class occur among the designs with the symmetric difference property; these designs have been characterized by Dillon and Schatz [25] and they have the minimum 2-rank, namely $2m + 2$. In this case there is not, to our knowledge, a rigidity theorem characterizing the "best" design[5] among those with the symmetric difference property and, in fact, the binary code of a design with the symmetric difference property is a complete invariant for those designs. The Higman design probably is characterized by its 2-rank, namely 22, but proving that appears to be rather difficult. Zvonimir Janko [35] has uncovered another $(176, 50, 14)$ design that is closely related to the Higman design and it has higher 2-rank. For a short discussion of all these matters the reader may wish to consult the paper by Parker and Tonchev, [50].[6]

For geometric designs it is well known that the codes, to be of interest, need to be over a prime field whose characteristic is that of the field of the geometry: see [3]. The codes of these designs are the Reed-Muller codes, $\mathcal{R}(r, m)$, in case $F = \mathbf{F}_2$, or various generalized Reed-Muller codes, as was proved mostly by the work of Delsarte et al. [20, 21, 22, 30]. These results are described in [3, Chapter 5] and in [5].

Rigidity theorems have been established: for Steiner triple systems (see Section 7) and for some classes of Hadamard designs (see Section 5). The negative answer to Question 2 described in Section 5 shows that a rigidity theorem similar to the results for Steiner triple systems (Theorems 14, 15 and 16) will not be found for Hadamard 3-$(2^m, 2^{m-1}, 2^{m-2}-1)$ designs, although we do have the Hamada-Ohmori rigidity theorem (Corollary 7.5.1 in [3]):

THEOREM 17 *The binary code of a* 3-$(2^m, 2^{m-1}, 2^{m-2}-1)$ *design has dimension at least* $m+1$ *and has at least* $2^{m+1}-2$ *vectors of weight* 2^{m-1}. *If either equality holds so does the other and this occurs if and only if the design is the design of points and hyperplanes in the affine geometry of dimension* m *over* $\mathbf{F}_2$.

9. Dual structures and the "point" code of a design

Bridges, Hall and Hayden in [10] gave some attention to the linear code generated by the transpose of an incidence matrix of a design, but they, like most others, thought that only those codes over fields whose characteristic divided the order of the design were of interest. It was Tonchev [68] who pointed out that these codes were of interest even when the prime did *not* divide the order of the design and, in an interesting paper, Tonchev and Weishaar [69] examined precisely these codes for all 80 Steiner triple systems on 15 points and found that the 80 binary codes of block length 35 were all different — in contrast to the usual codes of block length 15, where there are only five different codes, one for each dimension k with $11 \leq k \leq 15$. For lack of a better name we call such a code the **point code** of the design; it is merely the code of the dual incidence structure — which is only a design in the symmetric case. The point code of the design of points and lines of $PG_3(\mathbf{F}_2)$ distinguishes itself from the other 79 codes not only by its dimension but also by the fact that it is the only one of these codes that has no vectors of weight 8 or 11 (and thus none of weight 24 or 27, since $\jmath$ is in the code); it has minimum weight 7, but there are other codes with this minimum weight. Moreover, the code is *not* contained in each of the other 79 (although it is contained in some of the 79), again in contrast to the usual codes: see Theorem 14. There are, in fact, numerous containment relations among the 80 codes, all of which were described in [69]. This was made possible computationally by the following well-known result [67, Exercise 1.1.17, page 7] — which is not normally stated in coding-theoretic terms, as below:

PROPOSITION 3 *Let* $0 < \lambda < r$ *and suppose given* v *vectors in* $\mathbf{F}_2^n$*, each of weight* r *and any two at distance* $2(r - \lambda)$*. Then,*

$$n \geq \frac{r^2 v}{r + \lambda(v-1)}$$

with equality if and only if the v *vectors form the columns of an incidence matrix of a* 2-(v, k, λ) *design, where* $k = \frac{r+\lambda(v-1)}{r}$.

Although the proposition is well-known its importance does not seem to have been recognized. It does, for example, show that a 2-design is the solution to an extremal problem in coding theory, a fact that went unobserved in [3].

Kevin Phelps [56] has conjectured that the binary point code of a Steiner triple system is a complete invariant; i.e. that two such point codes are isomorphic if and only if the triple systems are. Even for the 80 triple systems on 15 points the matter is rather delicate since some of the codes have the same weight distributions. In the only other case so far considered, the two triple systems on 13 points, the two binary point codes have different weight distributions; in each case they are $[26, 13]$ codes since the order of the designs is 5. Because the order of these designs is 5, rank considerations alone show that the two designs have the same codes, in the usual sense, for *every* prime p.

Of course, in general, the binary point codes of a class of designs with the same parameters cannot characterize the designs since there are symmetric designs for which this is patently false (for example, the four projective planes of order nine) and non-symmetric examples are also available. Curiously, however, the 17 embeddable unitals on 28 points [54] are so characterized [59]; we do not know what the "correct", most general conjecture might be — if, in fact, Phelps's conjecture for the triple systems is true — and the subject probably deserves further investigation. It is even theoretically possible to have a point code of one design contain the point code of another design with different parameters, but we lack an example of this. As far as we know there has been nothing substantial done with point codes except over the binary field (but see [50]).

10. Strongly regular graphs and their p-ranks

Although codes associated with graphs have not been extensively studied, some significant results concerning the p-rank of adjacency matrices of strongly regular graphs have recently been established: see Brouwer and van Eijl [12] and Peeters [52]. Since these concepts are closely related to certain constructions we have examined for designs, we will describe here some of the new results.[7]

Recall that a strongly regular graph Γ with parameters (v, k, λ, μ) is a regular undirected graph on v vertices with valency k, such that λ is the number of vertices adjacent to both of two adjacent vertices and μ the number of vertices adjacent to both of two non-adjacent vertices. An adjacency matrix[8] for Γ is also an incidence matrix for a quasi-symmetric

1-design (see Shrikhande and Sane [62]) on the v points with block size k. This quasi-symmetric design is a symmetric (v, k, λ) design if and only if $\lambda = \mu$.

Let Γ be a strongly regular graph with parameters (v, k, λ, μ), and let A be an adjacency matrix for Γ. It is well known (see, for example, Cameron and van Lint [14, Chapter 2]) that A has three eigenvalues, namely k and two others, ρ and σ, with

$$\rho, \sigma = \frac{1}{2}(\lambda - \mu \pm \sqrt{(\lambda - \mu)^2 + 4(k - \mu)}).$$

Calling, as usual, the multiplicities f and g we have that

$$\begin{aligned} v &= f + g + 1, \\ 0 &= k + f\rho + g\sigma, \end{aligned}$$

and that $\lambda - \mu = \rho + \sigma$ and $\mu - k = \rho\sigma$. From this it follows that

$$f, g = \frac{1}{2}(v - 1 \pm \frac{(v-1)(\mu - \lambda) - 2k}{\sqrt{(\mu - \lambda)^2 + 4(k - \mu)}}),$$

respectively. We refer to the case $f = g$ as the **half** case. In the half case the parameters are $(4t + 1, 2t, t - 1, t)$, for some $t \geq 1$. The following is a consequence of results proved in [12]:

THEOREM 18 *If p is a prime and c is any integer, then the p-rank of the matrix $A + cI_v$, where A is an adjacency matrix of a strongly regular graph Γ with parameters (v, k, λ, μ), is completely determined by these parameters except possibly for the p-ranks of*

- *$A + \frac{1}{2}(p+1)I_v$ with p dividing v in the half case, and*
- *$A - \sigma I_v$ with p dividing $(\rho - \sigma)$ where ρ and σ are the eigenvalues of A other than k and are assumed to be integral,*

in which case the p-rank is upper bounded by $\min\{f + 1, g + 1\}$ *where f and g are the multiplicities of ρ and σ.*

Notice that if we have a strongly regular graph from a Steiner 2-$(v, k, 1)$ design $\mathcal{D}$, then $\rho = r - 1 - k$, $\sigma = -k$, $f = v - 1$ and $g = b - v$. Furthermore, if M is an incidence matrix for $\mathcal{D}$ and A the adjacency matrix for the block graph Γ of $\mathcal{D}$ using the same ordering of blocks, then

$$MM^t = A + kI_b,$$

and

$$\text{rank}_F(M) = \text{rank}_F(\mathcal{D}) = \text{rank}_F(A + kI_b) + \dim(\text{Hull}_F(\mathcal{D})) \quad (1)$$

over any field F.

In [52], Peeters's aim is to characterize strongly regular graphs by their parameters *and* the p-rank for one of the values of p that fall into the exclusions of the theorem (called the "relevant" p-ranks of Γ). In particular, for the half case he shows

THEOREM 19 *Let Γ be a strongly regular graph with parameters $(4t+1, 2t, t-1, t)$ and adjacency matrix A. If p divides v but p^2 does not, then $A + \frac{1}{2}(p+1)I_v$ has p-rank $2t+1$.*

The block graphs of Steiner triple systems are examined by Peeters in [51]; the theorem of Doyen et al. (Theorem 13) gives the p-ranks of the Steiner 2-$(v, 3, 1)$ designs, and the above equality gives the rank of $A + 3I_b$ where A is an adjacency matrix for the block graph. Hence when $p \geq 5$ the p-rank of $A + 3I_b$ is v; if $p = 2$ then from [40, Proposition 2] we know that $\mathrm{Hull}_2(\mathcal{D}) = C^\perp$ if 2 divides the order $n = (v-3)/2$ (and clearly also if not), so that equation (1) above gives the 2-rank of $A + 3I_b$ as $2 \times \mathrm{rank}_2(\mathcal{D}) - v$. For $p = 3$, if 3 does not divide the order n then also 3 does not divide v, and so $\mathrm{rank}_3(\mathcal{D}) = v - 1$ and $\mathrm{Hull}_3(\mathcal{D}) = \{0\}$, so that $A + 3I_b$ has 3-rank $v - 1$. If 3 divides n then the same result from [40] shows that $\mathrm{Hull}_3(\mathcal{D}) = C^\perp$, and the 3-rank of $A + 3I_b$ as $2 \times \mathrm{rank}_3(\mathcal{D}) - v$, except in the case where 9 does not divide v and $\mathcal{D}$ has exactly three affine hyperplanes (i.e. $d_A = 1$), in which case it follows that $\mathrm{Hull}_3(\mathcal{D}) = \langle \jmath \rangle$ and that the 3-rank of $A + 3I_b$ is $v - 3$. (Note that there is an error in [40, Proposition 2] in the statement for $p = 3$ in that it is not true that $C_3(\mathcal{D})^\perp \subseteq C_3(\mathcal{D})$ in the case where 9 does not divide v and $\mathcal{D}$ has exactly three affine hyperplanes.)

For example, the Steiner triple systems on 15 points give block graphs on 35 points and matrices $A + I_{35}$ with 2-rank in the set $\{7, 9, 11, 13, 15\}$. By way of an exercise, we used Magma to obtain the weight distribution of the code of dimension 7 spanned over $\mathbf{F}_2$ by the matrix $A + I_{35}$ for the 2-(15,3,1) design of points and lines of $PG_3(\mathbf{F}_2)$:

$$< 0, 1 >, < 15, 28 >, < 16, 35 >, < 19, 35 >, < 20, 28 >, < 35, 1 >$$

and its hull:

$$< 0, 1 >, < 16, 35 >, < 20, 28 >$$

which is the even-weight subcode in this case. The notation here is that used in Magma: the pair $< i, A_i >$ denotes A_i vectors of weight i. Similar computations could be made for the other 79 designs: we do not know if they give distinct codes or not, as in the case of the dual structures mentioned in Section 9.

We mention again that not a great deal seems to be known about the codes spanned by these matrices. Codes associated with quasi-symmetric designs have been examined by various authors: see Shrikhande and Sane [62, Chapter X] for some of the available results. Even in the case of an adjacency matrix giving a symmetric design the code of the design is not well documented. Note, for example, that it was the hermitian unital on 28 points (with $k = 4$) that gave a block graph yielding a code that provided a negative answer to Question 2.

Acknowledgments

Both authors would like to thank René Peeters and John Dillon for a careful reading of an earlier draft of this article and saving us from blunders. We would also like to respectively thank INRIA and the University of Nebraska - Lincoln for their hospitality during the preparation of parts of the manuscript.

Notes

1. Ovals are also called *hyperovals* in the literature.
2. A *constant* vector is a non-zero vector in which all entries are either 0 or a where a is some non-zero field element.
3. The Ree unital on 28 points has been recently characterized by its 2-rank by G. McGuire, H. Ward, and V. Tonchev. (Note added in proof by the editor, June, 1996).
4. The perfect code will always be non-linear unless the system is the classical one of points and lines of $PG_{m-1}(\mathbf{F}_2)$.
5. Namely the one with the doubly-transitive automorphism group.
6. In fact, the subject of this work is the construction of the Higman design via coding theory, utilizing the binary point codes of the derived or residual designs, but a discussion of all symmetric designs possessing a doubly-transitive automorphism group is included.
7. See also the last paragraph of this section for a further reference.
8. An *adjacency matrix* of a graph is a square zero-one matrix whose rows and columns are both indexed by vertices with a 1 in a given position if the respective vertices are adjacent; it is symmetric if the rows and columns are indexed in the same order and, in this case, has 0s on the diagonal. If the graph is strongly regular any row has k ones and any two distinct rows have either λ or μ ones in common.

References

1. Bruno Ratsimandefitra Andriamanalimanana. *Ovals, Unitals and Codes.* PhD thesis, Lehigh University, 1979.
2. E. F. Assmus, Jr. and J. D. Key. Arcs and ovals in the hermitian and Ree unitals. *European J. Combin.*, 10:297–308, 1989.
3. E. F. Assmus, Jr. and J. D. Key. *Designs and their Codes.* Cambridge University Press, 1992. Cambridge Tracts in Mathematics, Vol. 103 (Second printing with corrections, 1993).
4. E. F. Assmus, Jr. and H. F. Mattson, Jr. On tactical configurations and error-correcting codes. *J. Combin. Theory*, 2:243–257, 1967.
5. Edward F. Assmus, Jr. and Jennifer D. Key. Codes and finite geometries. Technical report, INRIA, 1993. Report No. 2027.
6. Sunanda Bagchi and Bhaskar Bagchi. Designs from pairs of finite fields: I. A cyclic unital $U(6)$ and other regular Steiner 2-designs. *J. Combin. Theory, Ser. A*, 52:51–61, 1989.
7. Aart Blokhuis and G. Eric Moorhouse. Some p-ranks related to orthogonal spaces. *J. Algebraic Combin.* To appear.
8. R. C. Bose and S. S. Shrikhande. On the construction of sets of mutually orthogonal latin squares and the falsity of a conjecture of Euler. *Trans. Amer. Math. Soc.*, 95:191–209, 1960.
9. W. Bosma and J. Cannon. *Handbook of Cayley Functions.* Department of Mathematics, University of Sydney, January 1993.
10. W. G. Bridges, M. Hall, Jr., and J. L. Hayden. Codes and designs. *J. Combin. Theory, Ser. A*, 31:155–174, 1981.
11. A. E. Brouwer. Some unitals on 28 points and their embeddings in projective planes of order 9. In M. Aigner and D. Jungnickel, editors, *Geometries and Groups*, pages 183–188. Springer-Verlag, 1981. Lecture Notes in Mathematics, No. 893.
12. A. E. Brouwer and C. J. van Eijl. On the p-rank of the adjacency matrices of strongly regular graphs. *J. Algebraic Combin.*, 1:329–346, 1992.
13. F. Buekenhout, A. Delandtsheer, and J. Doyen. Finite linear spaces with flag-transitive groups. *J. Combin. Theory, Ser. A*, 49:268–293, 1988.
14. P. J. Cameron and J. H. van Lint. *Designs, Graphs, Codes and their Links.* Cambridge: Cambridge University Press, 1991. London Mathematical Society Student Texts 22.
15. John Cannon and Catherine Playoust. *An Introduction to Magma.* School of Mathematics and Statistics, University of Sydney, 1994.

16. L. L. Carpenter. Oval designs in desarguesian projective planes. *Des. Codes Cryptogr.*, 1995. To appear.
17. L. L. Carpenter and J. D. Key. Reed-Muller codes and Hadamard designs from ovals. *J. Combin. Math. Combin. Comput.*, 1995. To appear.
18. James A. Davis and Jonathan Jedwab. A summary of Menon difference sets. *Congressus Numerant.*, 93:203–207, 1993.
19. Michel Dehon. Ranks of incidence matrices of t-designs $S_\lambda(t, t+1, \lambda)$. *European J. Combin.*, 1:97–100, 1980.
20. P. Delsarte, J. M. Goethals, and F. J. MacWilliams. On generalized Reed-Muller codes and their relatives. *Inform. and Control*, 16:403–442, 1970.
21. Philippe Delsarte. A geometric approach to a class of cyclic codes. *J. Combin. Theory*, 6:340–358, 1969.
22. Philippe Delsarte. On cyclic codes that are invariant under the general linear group. *IEEE Trans. Inform. Theory*, 16:760–769, 1970.
23. Immo Diener, Eberhard Schmitt, and Hans Ludwig de Vries. All 80 Steiner triple systems on 15 points are extendable. *Discrete Math.*, 55:13–19, 1985.
24. J. F. Dillon. Private communication.
25. J. F. Dillon and J. R. Schatz. Block designs with the symmetric difference property. In Robert L. Ward, editor, *Proceedings of the NSA Mathematical Sciences Meetings*, pages 159–164. The United States Government, 1987.
26. Steven Dougherty. Nets and their codes. *Des. Codes Cryptogr.*, 3:315–331, 1993.
27. Jean Doyen, Xavier Hubaut, and Monique Vandensavel. Ranks of incidence matrices of Steiner triple systems. *Math. Z.*, 163:251–259, 1978.
28. Tuvi Etzion and Alexander Vardy. Perfect binary codes: constructions, properties and enumeration. *IEEE Trans. Inform. Theory*, 40:754–763, 1994.
29. Meinolf Geck. Irreducible Brauer characters of the 3-dimensional special unitary groups in non-defining characteristic. *Comm. Algebra*, 18:563–584, 1990.
30. Jean-Marie Goethals and Philippe Delsarte. On a class of majority-logic decodable cyclic codes. *IEEE Trans. Inform. Theory*, 14:182–188, 1968.
31. N. Hamada. The geometric structure and the p-rank of an affine triple system derived from a nonassociative Moufang loop with the maximum associative center. *J. Combin. Theory, Ser. A*, 30:285–297, 1981.
32. Gerhard Hiss. Private communication.
33. G. Hölz. Construction of designs which contain a unital. *Arch. Math.*, 37:179–183, 1981.
34. Wen-Ai Jackson. A characterization of Hadamard designs with $SL(2, q)$ acting transitively. *Geom. Dedicata*, 46:197–206, 1993.
35. Zvonimir Janko. On symmetric designs with parameters $(176, 50, 14)$. Preprint.
36. William M. Kantor. Classification of 2-transitive symmetric designs. *Graphs Combin.*, 1:165–166, 1985.
37. J. D. Key. Ternary codes of Steiner triple systems. *J. Combinatorial Designs*, 2:25–30, 1994.
38. J. D. Key and F. E. Sullivan. Steiner systems from binary codes. Submitted.
39. J. D. Key and F. E. Sullivan. Steiner triple systems with many affine hyperplanes. Submitted
40. J. D. Key and F. E. Sullivan. Codes of Steiner triple and quadruple systems. *Des. Codes Cryptogr.*, 3:117–125, 1993.
41. Kirsten Mackenzie. *Codes of Designs*. PhD thesis, University of Birmingham, 1989.
42. A. Maschietti. Hyperovals and Hadamard designs. *J. Geom.*, 44:107–116, 1992.
43. R. Mathon. Constructions of cyclic Steiner 2-designs. *Ann. Discrete Math.*, 34:353–362, 1987.
44. Rudolf Mathon and Gordon F. Royle. The translation planes of order 49. *Des. Codes Cryptogr.*, 5:57–72, 1995.
45. N. S. Mendelsohn and Stephen H. Y. Hung. On the Steiner systems $S(3, 4, 14)$ and $S(4, 5, 15)$. *Utilitas Math.*, 1:5–95, 1972.
46. T. S. Michael. The p-ranks of skew Hadamard designs. *J. Combin. Theory, Ser. A*. To appear.
47. G. Eric Moorhouse. Bruck nets, codes, and characters of loops. *Des. Codes Cryptogr.*, 1:7–29, 1991.
48. Brian Mortimer. The modular permutation representations of the known doubly transitive groups. *Proc. London Math. Soc. (3)*, 41:1–20, 1980.
49. T. Norwood. Private communication.
50. Christopher Parker and Vladimir D. Tonchev. Linear codes and doubly-transitive symmetric designs. Preprint.

51. René Peeters. On the p-ranks of the block graphs of Steiner triple systems. Preprint.
52. René Peeters. Uniqueness of strongly regular graphs having minimal p-rank. Tilburg University, Department of Economics Research Memorandum, FEW 626.
53. T. Penttila and I. Pinneri. Irregular hyperovals in $PG(2,4)$. *J. Geom.*, 51:89–100, 1994.
54. Tim Penttila and Gordon F. Royle. Sets of type (m,n) in affine and projective planes of order nine. To appear: *Design Codes and Cryptography.*
55. Tim Penttila, Gordon F. Royle, and M. K. Simpson. Hyperovals in the known projective planes of order 16. In preparation.
56. K. T. Phelps. Private communication.
57. K. T. Phelps. A combinatorial construction of perfect codes. *SIAM J. Alg. Disc. Meth.*, 4:398–403, 1983.
58. Alexander Pott. On abelian difference set codes. *Des. Codes Cryptogr.*, 2:263–271, 1992.
59. Gordon F. Royle. Private communication.
60. Beniamino Segre. Ovals in a finite projective plane. *Canad. J. Math.*, 7:414–416, 1955.
61. F.D. Shobe. *On a class of Steiner systems and their codes.* PhD thesis, Clemson University, 1995. Submitted.
62. Mohan S. Shrikhande and Sharad S. Sane. *Quasi-Symmetric Designs.* Cambridge University Press, 1991. London Mathematical Society Lecture Notes Series 164.
63. Deirdre Langacher Smeltzer. *Topics in difference sets in 2-groups.* PhD thesis, University of Virginia, 1994.
64. J. Steiner. Combinatorische Aufgabe. *J. Reine Angew. Math.*, 45:181–182, 1853.
65. Luc Teirlinck. On projective and affine hyperplanes. *J. Combin. Theory, Ser. A*, 28:290–306, 1980.
66. Vladimir D. Tonchev. Quasi-symmetric 2-(31,7,7) designs and a revision of Hamada's conjecture. *J. Combin. Theory, Ser. A*, 42:104–110, 1986.
67. Vladimir D. Tonchev. *Combinatorial Configurations Designs, Codes, Graphs.* Pitman Monographs and Surveys in Pure and Applied Mathematics, No. 40. New York: Longman, 1988. Translated from the Bulgarian by Robert A. Melter.
68. Vladimir D. Tonchev. Quasi-symmetric designs, codes, quadrics, and hyperplane sections. *Geom. Dedicata*, 48:295–308, 1993.
69. Vladimir D. Tonchev and Robert S. Weishaar. Steiner triple systems of order 15 and their codes. *J. Statist. Plann. Inference.* To appear.
70. Michael A. Wertheimer. Oval designs in quadrics. *Contemp. Math.*, 111:287–297, 1990. Published by the American Mathematical Society.
71. Richard M. Wilson. Nonisomorphic Steiner triple systems. *Math. Z.*, 135:303–313, 1974.
72. M. -y. Xia. Some infinite classes of special Williamson matrices and difference sets. *J. Combin. Theory, Ser. A*, 61:230–242, 1992.

Designs, Codes and Cryptography, 9, 29–38 (1996)
© 1996 Kluwer Academic Publishers, Boston. Manufactured in The Netherlands.

Some t-Homogeneous Sets of Permutations

JÜRGEN BIERBRAUER, STEPHEN BLACK, AND YVES EDEL
Department of Mathematical Sciences, Michigan Technological University, Houghton, MI 49931 (USA)
IBM Heidelberg (Germany)
Mathematisches Institut der Universität, Im Neuenheimer Feld 288, 69120 Heidelberg (Germany)

Communicated by: D. Jungnickel

Received October 26, 1994; Accepted February 15, 1996

Abstract. **Perpendicular Arrays** are ordered combinatorial structures, which recently have found applications in cryptography. A fundamental construction uses as ingredients **combinatorial designs** and **uniformly t-homogeneous sets of permutations**. We study the latter type of objects. These may also be viewed as generalizations of t-homogeneous groups of permutations. Several construction techniques are given. Here we concentrate on the **optimal** case, where the number of permutations attains the lower bound. We obtain several new optimal such sets of permutations. Each example allows the construction of infinite families of perpendicular arrays.

Keywords: Permutation sets, permutation groups, perpendicular arrays, authentication

1. Introduction

DEFINITION 1 *A* **perpendicular array** $PA_\lambda(t, k, v)$ *is a multiset $\mathcal{A}$ of injective mappings from a k-set C into a v-set E, which satisfies the following:*

- *for every t-subset $U \subseteq C$ and every t-subset $W \subseteq E$ the number of elements of $\mathcal{A}$ (eventually counted with multiplicities) mapping U onto W is λ, independent of the choice of U and W.*

Alternatively $\mathcal{A}$ may be viewed as an array with C as set of columns and E as set of entries, where each mapping contributes a row. Here we are primarily interested in the case $k = v = n$. A $PA_\mu(t, n, n)$ may be described as a **μ-uniform t-homogeneous multiset of permutations on n objects.** We speak of a $PA(t, n, n)$ if we are not interested in the value of μ. A $PA(t, n, n)$ is **inductive**, equivalently is an $APA(t, n, n)$ if it is a $PA(w, n, n)$ for every w, $1 \leq w \leq t$. Every $PA(t, n, n)$ is inductive provided $t \leq (n + 1)/2$ (see [8]). In the above APA stands for **authentication** perpendicular array. This term was coined by D. R. Stinson ([8]) and further generalized in [2]. The notation stems from an application in the cryptographical theory of unconditional secrecy and authentication. The general definition is as follows:

DEFINITION 2 *An* **authentication perpendicular array** $APA_{\mu}(t,k,v)$ *is a* $PA_{\mu}(t,k,v)$ *which satisfies in addition*

- *For any* $t' < t$, *and for any* $t'+1$ *distinct entries we have, that among all the rows of the array* A *which contain all those entries, any subset of* t' *of those entries occurs in all possible subsets of* t' *columns equally often.*

Thus PA and APA may be viewed as t-designs, where the blocks are ordered. The basic ingredients in the construction of general APA and related structures are

- t-designs, and
- $APA(t,n,n)$.

In fact the unordered structure underlying an $APA(t,k,v)$ is a t-design with block-size k. An $APA(t,k,k)$ may be used to yield the required ordered structure. (see [8]).

In the sequel we concentrate on sets (instead of multisets) of permutations. Such arrays may be called **simple.**

Examples of $APA(t,n,n)$ are furnished by t-homogeneous **groups** of permutations. However, as a consequence of the characterization of finite simple groups all the t-homogenous groups of permutations are known ($2 \le t \le (n+1)/2$). Aside from the alternating and symmetric groups there is no infinite family of t-homogeneous groups on n objects when $3 < t \le (n+1)/2$. It is therefore necessary to find different methods of constructing $APA_{\mu}(t,n,n)$. Given t and n we consider the problem of constructing $APA_{\mu}(t,n,n)$ which are as small as possible. This is equivalent to minimizing μ. As the number of permutations of an $APA_{\mu}(t,n,n)$ is divisible by $\binom{n}{w}$ for every w, $1 \le w \le t$, it follows that μ is divisible by $LCM\{\binom{n}{w} \,|w = 1,2,\ldots t)\}/\binom{n}{t}$.

DEFINITION 3 *Put*

$$\mu_0(t,n) = LCM\left\{\binom{n}{w} |w = 1,2,\ldots t)\right\}/\binom{n}{t}.$$

An $APA_{\mu}(t,n,n)$ *is called* **optimal** *if* $\mu = \mu_0(t,n)$.

We list the values of this function for small t :

$$\mu_0(1,n) = 1.$$

$$\mu_0(2,n) = \begin{cases} 1 & \text{if } n \text{ odd} \\ 2 & \text{if } n \text{ even.} \end{cases}$$

$$\mu_0(3,n) = \begin{cases} 1 & \text{if } n \equiv 2 (mod\ 3) \\ 3 & \text{otherwise.} \end{cases}$$

$$\mu_0(4,n) = \begin{cases} 1 & \text{if } n \equiv 3, 11 (mod\ 12) \\ 2 & \text{if } n \equiv 5, 9 (mod\ 12) \\ 3 & \text{if } n \equiv 7 (mod\ 12) \\ 4 & \text{if } n \equiv 0, 2, 6, 8 (mod\ 12) \\ 6 & \text{if } n \equiv 1 (mod\ 12) \\ 12 & \text{if } n \equiv 4, 10 (mod\ 12). \end{cases}$$

Our primary interest here is in the construction of optimal $APA(t, n, n)$. We may restrict attention to the case $t \leq (n+1)/2$. This is due to the fact that a uniformly t-homogeneous set of permutations on n objects is also uniformly $(n-t)$-homogeneous.

For $t = 1$ there is no problem. An $APA_1(1, n, n)$ is nothing but a latin square of order n. For $t = 2$ and $n = q$ a prime-power, the affine group $AGL_1(q)$ is an $APA_2(2, q, q)$. This is optimal if q is a power of 2. If q is odd, then $AGL_1(q)$ contains an $APA_1(2, q, q)$ (see [7]). The projective group $PSL_2(q)$ is an $APA_3(3, q+1, q+1)$ if q is a prime-power, $q \equiv 3(mod\ 4)$. This is optimal if $q \equiv 3, 11(mod\ 12)$. This yields optimal

$$APA_3(3, 12, 12), APA_3(3, 24, 24), APA_3(3, 28, 28), \ldots.$$

These are the only known infinite families of optimal $APA(t, n, n)$. In [5] an $APA_2(2, 6, 6)$ was constructed. In [3] it was shown that the group $PSL_2(q), q \not\equiv 3(mod\ 4)$, can be **halved** as a uniformly 2-homogeneous set of permutations on the projective line. The case $q = 5$ yields another construction of an $APA_2(2, 6, 6)$. An $APA_3(3, 6, 6)$ is constructed in [6] and [1]. A recursive construction given in [2], Corollary 6 when applied to an $APA_1(2, 5, 5)$ (equivalently: an $APA_1(3, 5, 5)$) also yields $APA_3(3, 6, 6)$.

The affine group $AGL_1(8)$ is an $APA_1(3, 8, 8)$, the group $A\Gamma L_1(32)$ is an $APA_1(3, 32, 32)$. An $APA_3(3, 9, 9)$ was constructed in [5] as a subset of the group $PGL_2(8)$. To the best of our knowledge these are all the optimal $PA(t, n, n)$, $t \leq (n+1)/2$ which have been known that far.

In Sections 2 and 3 we describe new methods of construction. Our main result is the following:

THEOREM 1

- *There exist (optimal)*
  - $APA_2(2, 10, 10)$
  - $APA_2(2, 12, 12)$
  - $APA_3(3, 7, 7)$
  - $APA_4(4, 8, 8)$
- *There is a (non-optimal) $APA_4(3, 11, 11)$ contained in the Mathieu group M_{11}.*
- *For $q \in \{3, 5, 7, 9\}$ the group $P\Gamma L_2(q^2)$ contains an $APA_{q-1}(2, q^2+1, q^2+1)$.*

The construction of optimal $APA(\lfloor n/2 \rfloor, n, n)$ is one of the central problems in the area.

The authors are convinced that this is a very hard problem in general. It is obvious that an optimal $APA(\lfloor n/2 \rfloor, n, n)$ is also an optimal $APA(t, n, n)$ for every t, $\lfloor n/2 \rfloor \leq t \leq n$. We get:

COROLLARY 1 *There exist (optimal)*

$$APA_3(4,7,7), APA_5(5,7,7), APA_{15}(6,7,7), APA_{105}(7,7,7),$$

$$APA_5(5,8,8), APA_{10}(6,8,8), APA_{35}(7,8,8), APA_{280}(8,8,8).$$

Moreover a symmetry in the construction yields the following corollary:

COROLLARY 2 *There exist (optimal)*

- $APA_2(2,5,6)$
- $APA_2(2,9,10)$
- $APA_2(2,11,12)$

2. The Double Coset-Method

DEFINITION 4 *Let G and H be subgroups of the symmetric group on n letters. A multiset $\mathcal{A}$ of permutations of the ground set is (G, H)-**admissible** if for every $g \in G, h \in H, \sigma \in \mathcal{A}$ we have $g\sigma h \in \mathcal{A}$ (if $\mathcal{A}$ is not simple we demand that the multiplicity of σ and of $g\sigma h$ are the same).*

Let now $\mathcal{A}$ be an $APA(t, n, n)$. For arbitrary permutations g and h the multiset $g\mathcal{A}h$ is an $APA(t, n, n)$ again. Therefore the set $G = \{g | g\mathcal{A} = \mathcal{A}\}$ is a group, the stabilizer of $\mathcal{A}$ under the action of the symmetric group S_n from the left. By operation from the right the situation is analogous. If $\mathcal{A}$ is (G, H)-admissible and α, β are arbitrary permutations of the ground set, then $\alpha\mathcal{A}\beta$ is $(\alpha G\alpha^{-1}, \beta^{-1}H\beta)$–admissible. We may therefore replace G and H by conjugate subgroups. If $\mathcal{A}$ is a (G, H)-admissible $APA_\mu(t, n, n)$, then the multiset $\mathcal{A}^{-1}$ of inverses is a (H, G)-admissible $APA_\mu(t, n, n)$. A (G, H)-admissible set of permutations may equivalently be described as a union of double cosets for G and H.

Let us visualize the multiset $\mathcal{A}$ of permutations as an array with n columns, where each element of $\mathcal{A}$, eventually counted with multiplicities, contributes a row, each row being a permutation. If $\mathcal{A}$ is (G, H)-admissible, then let H operate on the set of columns, whereas G permutes the entries of the array.

Consider first the problem of constructing $APA_2(2, n, n), n$ even. Such an array $\mathcal{A}$ has $n(n-1)$ elements. It is then conceivable that $\mathcal{A}$ is (G, G)-admissible, where G is a group of order $n-1$. Assume $G = Z_{n-1}$ in its natural action on n points, $G = \langle \zeta \rangle, \zeta =$

$(\infty)(0, 1, 2, \ldots n-2)$. Then $\mathcal{A}$ must be the union of two double cosets, one of which is Z_{n-1} itself:

$$\mathcal{A} = Z_{n-1} \cup Z_{n-1} \cdot \sigma_0 \cdot Z_{n-1}.$$

Thus $\mathcal{A}$ is determined by one permutation σ_0. Observe that σ_0 may be replaced by an arbitrary element of the same double coset. As $\mu = 2$, there must be an element in $Z_{n-1} \cdot \sigma_0 \cdot Z_{n-1}$ fixing the set $\{\infty, 0\}$. As $\mathcal{A}$ is an $APA_{n-1}(1, n, n)$, no element of $\mathcal{A} - Z_{n-1}$ can fix ∞. We choose σ_0 to be the unique element of $\mathcal{A}$ affording the operation $\sigma_0 : \infty \longleftrightarrow 0$. Write $\sigma_0 = (\infty, 0) \cdot \rho_0$, where ρ_0 is a permutation of $\{1, 2, \ldots n-2\}$.

Consider the circle $C = C_{n-1}$ of length $n-1$ with set $\{0, 1, 2, \ldots n-2\}$ of vertices and neighbourhood relation

$$i \perp j \iff |i - j| \equiv 1 (mod\ n-1).$$

Let $d(\ ,\ )$ denote the distance in C, $\Delta = \{1, 2, \ldots \frac{n}{2} - 1\}$ the set of distances $\neq 0$. For every $\delta \in \Delta$ let P_δ be the set of unordered pairs $\{x, y\}$ of vertices of C satisfying $xy \neq 0, d(x, y) = \delta$. Observe that $|P_\delta| = n - 3$ for every $\delta \in \Delta$.

THEOREM 2 *Let n be an even number. Then the following are equivalent:*

- *There is a (Z_{n-1}, Z_{n-1})-admissible $APA_2(2, n, n)$.*
- *There is a permutation ρ of $\{0, 1, 2, \ldots n-2\}$, $\rho(0) = 0$ such that for every $\delta \in \Delta$ the following is satisfied:*

$$|\rho(P_\delta) \cap P_\delta| = 1.$$

$$|\rho(P_\delta) \cap P_{\delta'}| = 2\ (\delta' \in \Delta, \delta' \neq \delta).$$

Proof. Write $Z_{n-1} = \{z(i) | i = 0, 1, 2, \ldots n-2\}$, where

$$z(i) : \tau \longmapsto \tau + i\ (mod\ n-1).$$

Then the typical element $z(i)\sigma_0 z(j)$ of $\mathcal{A} - Z_{n-1}$ affords the operation

$$\tau \longmapsto (\tau + i)^{\sigma_0} + j.$$

Let A, B be two unordered pairs of elements in $\{\infty, 0, 1, 2, \ldots, n-2\}$. We have to make sure that exactly two elements of $\mathcal{A}$ map A onto B. We have

$$z(l - j) : \infty \longrightarrow \infty, j \longrightarrow l.$$

$$z(-j)\sigma_0 z(l) : j \longrightarrow \infty \longrightarrow l.$$

$$z((l-k)^{\sigma_0^{-1}} - j)\sigma_0 z(k) : \infty \longrightarrow k, j \longrightarrow l.$$

$$z(-i)\sigma_0 z(l-(j-i)^{\sigma_0}) : i \longrightarrow \infty, j \longrightarrow l.$$

In fact the element of $\mathcal{A}$ affording one of these operations is uniquely determined in each case. This shows that the condition is satisfied whenever $\infty \in A$ or $\infty \in B$, independent of the choice of ρ_0.

Let now $A = \{i, j\}$, $B = \{k, l\}$, where $\infty \notin A \cup B, i \neq j, k \neq l$. Exactly then is there an element of Z_{n-1} mapping A onto B if $d(i, j) = d(k, l)$. This element is then uniquely determined. An element $z(\alpha)\sigma_0 z(\beta)$ affords the operation $i \mapsto k$, $j \mapsto l$ if and only if

$$(i+\alpha)^{\rho_0} + \beta = k$$

$$(j+\alpha)^{\rho_0} + \beta = l$$

The condition on α is $(i+\alpha)^{\rho_0} - (j+\alpha)^{\rho_0} = k - l$. Interchanging k and l we see that a necessary and sufficient condition for α is

$$d((i+\alpha)^{\rho_0}, (j+\alpha)^{\rho_0}) = d(k, l).$$

The Theorem is now obvious. ■

Thus the existence of a (Z_{n-1}, Z_{n-1})-admissible $APA_2(2, n, n)$ is equivalent to the existence of a permutation on $n-1$ letters, which fixes one letter and destroys the metric given by a circle of length $n-1$ in the most effective way.

THEOREM 3 *Let n be even. If n is a power of 2 or $n \in \{6, 12\}$, then there is a (Z_{n-1}, Z_{n-1})-admissible $APA_2(2, n, n)$.*

Proof. If $n = q$ is a power of 2, then the group $AGL_1(q)$ is an $APA_2(2, q, q)$. As it contains the multiplicative group of the field $\mathbb{F}_q$, it is (Z_{n-1}, Z_{n-1})-admissible.

For $n = 6$ and $n = 12$ it suffices, by the preceding theorem, to give the permutation ρ_0. If $n = 6$, then ρ_0 is uniquely determined: $\rho_0 = (1, 4)$. If $n = 12$, we may choose

$$\rho_0 \in \{\rho_1 = (1, 3, 9, 5, 4)(2, 8, 10, 7, 6), \rho_2 = \rho_1^{-1},$$

$$\rho_3 = (1, 7)(2, 5)(3, 10)(4, 6)(8, 9), \rho_4 = (1, 8)(2, 3)(4, 10)(5, 7)(6, 9)\}.$$

■

An exhaustive search showed that that for $n \in \{10, 14, 18, 20, 22\}$ there is no (Z_{n-1}, Z_{n-1})-admissible $APA_2(2, n, n)$.

DEFINITION 5 *Fix $Z = Z_{n-1}$ and $C = C_{n-1}$ as before. Let $\Pi = \Pi_{n-1}$ be the set of permutations ρ_0 such that $\rho = (0)\rho_0$ satisfies the conditions of Theorem 2.*

In fact $\Pi_5 = \{(1,4)\}$, $\Pi_{11} = \{\rho_1, \rho_1^{-1}, \rho_3, \rho_4\}$, where the permutations are given in the proof of the preceding Theorem.

LEMMA 1 *If $\rho \in \Pi$, then $I(\rho) \in \Pi$ and $N(\rho) \in \Pi$, where the involutory operations I and N are defined by*

$$I(\rho)(\tau) = \rho^{-1}(\tau) \tag{1}$$
$$N(\rho)(\tau) = \rho(-\tau). \tag{2}$$

Moreover the group $\langle I, N\rangle$ generated by I and N is dihedral of order 8.

Proof. This is a consequence of the following easily checked facts: I and N are involutory operations mapping Π onto itself. The product IN has order 4. ■

The elements of Π_{11} are rather interesting. We have

$$\rho_3(x) = \prod_{x \in \mathbb{F}_{11}^{*2}} (x, -4x),$$

$$\rho_1(x) = x \cdot 3^{\left(\frac{x}{11}\right)},$$

where $\left(\frac{a}{b}\right)$ is the Legendre symbol. We tried to generalize this to larger fields but were not successful. If $\mathcal{A} = \mathcal{A}(\rho_0) = Z_{n-1} \cup Z_{n-1} \cdot \rho_0 \cdot Z_{n-1}$ is an $APA_2(2, n, n)$, then $\mathcal{A}(\rho_0^{-1})$ is simply the set of inverses. In contrast to this the relation between $\mathcal{A}(\rho_0)$ and $\mathcal{A}(g(\rho_0))$ for other $g \in \langle I, N\rangle$ may be rather mysterious. It happens that one of them is sharply 2-transitive while the other is not. Even more can happen. Consider the case $n = 12$ again. The group $\langle I, N\rangle$ operates transitively on Π_{11}. In spite of that the group generated by $\mathcal{A}(\rho_1)$ (and by $\mathcal{A}(\rho_1^{-1})$) is the full symmetric group S_{12}, whereas $\mathcal{A}(\rho_3)$ and $\mathcal{A}(\rho_4)$ generate a copy of the Mathieu group M_{12}.

The following constructions of (G, H)-admissible sets of permutations are computer-results. They were obtained by the third author. In each case we give G (operating on the columns of the array), H (operating on the entries of the array) and the **generator-matrix,** whose rows are the generators of double-cosets. The set of symbols is $\{1, 2, \ldots, n\}$. It is easy to check that the arrays have the desired properties.

THEOREM 4 *Let $\mathcal{A}$ be a union of double cosets of groups G and H, where the double coset-representatives are the rows of the generator-matrix M.*

- *Let* $G = \langle (1,2,3)(4,5,6)(7,8,9), (1,4,7)(2,5,8)(3,6,9)\rangle$, $H = \langle (1,5,6,7,10)(2,4,9,3,8)\rangle$,

$$M = \begin{matrix} 1 & 2 & 3 & 4 & 5 & 6 & 7 & 8 & 9 & 10 \\ 1 & 4 & 9 & 6 & 8 & 2 & 5 & 10 & 7 & 3 \end{matrix}$$

Then $\mathcal{A}$ is an $APA_2(2, 10, 10)$.

- *Let* $G = \langle(1,2,3,4,5,6,7)\rangle$, $H = \langle(2,3,4,5,6)\rangle$,

$$M = \begin{array}{ccccccc} 1 & 2 & 3 & 4 & 6 & 5 & 7 \\ 1 & 2 & 3 & 5 & 7 & 4 & 6 \\ 1 & 2 & 4 & 3 & 6 & 7 & 5 \end{array}$$

Then $\mathcal{A}$ *is an* $APA_3(3,7,7)$.

- *Let* $G = \langle(2,3,4,5,6,7,8)\rangle$, $H = \langle(4,5,6,7,8)\rangle$,

$$M = \begin{array}{cccccccc} 1 & 2 & 3 & 4 & 5 & 6 & 7 & 8 \\ 2 & 1 & 4 & 3 & 5 & 6 & 8 & 7 \\ 2 & 5 & 1 & 3 & 4 & 6 & 8 & 7 \\ 2 & 4 & 6 & 1 & 3 & 5 & 8 & 7 \\ 2 & 6 & 3 & 1 & 4 & 7 & 8 & 5 \\ 2 & 7 & 8 & 1 & 4 & 6 & 3 & 5 \\ 2 & 8 & 4 & 1 & 6 & 3 & 5 & 7 \\ 2 & 8 & 6 & 1 & 4 & 7 & 3 & 5 \end{array}$$

Then $\mathcal{A}$ *is an* $APA_4(4,8,8)$.

- *Let* $G = \langle(1,2,3,4,5,6,7,8,9,10,11)\rangle$, $H = \langle(2,3,4,5,6)(7,8,9,10,11)\rangle$

$$M = \begin{array}{ccccccccccc} 1 & 2 & 4 & 6 & 3 & 11 & 9 & 7 & 8 & 5 & 10 \\ 1 & 2 & 6 & 7 & 10 & 11 & 8 & 5 & 3 & 4 & 9 \\ 1 & 2 & 11 & 5 & 3 & 10 & 4 & 8 & 6 & 7 & 9 \\ 1 & 2 & 4 & 10 & 6 & 5 & 11 & 9 & 3 & 7 & 8 \\ 1 & 2 & 10 & 9 & 8 & 5 & 3 & 7 & 6 & 4 & 11 \\ 1 & 4 & 8 & 5 & 10 & 3 & 2 & 11 & 7 & 6 & 9 \\ 1 & 2 & 10 & 5 & 4 & 6 & 8 & 9 & 7 & 11 & 3 \\ 1 & 7 & 6 & 5 & 4 & 11 & 2 & 8 & 9 & 3 & 10 \\ 1 & 2 & 8 & 3 & 10 & 6 & 9 & 4 & 5 & 7 & 11 \\ 1 & 5 & 8 & 6 & 9 & 4 & 2 & 7 & 10 & 3 & 11 \\ 1 & 2 & 7 & 6 & 4 & 9 & 5 & 8 & 3 & 10 & 11 \\ 1 & 7 & 9 & 4 & 10 & 5 & 2 & 6 & 3 & 8 & 11 \end{array}$$

Then $\mathcal{A}$ *is an* $APA_4(3,11,11)$.

Our construction of an $APA_2(2, 10, 10)$ will be generalized in the next section.

The second author found the first example of an $APA_2(2, 10, 10)$ in January 1992. His example is contained in the symmetric group S_6 in its 2-transitive action on 10 points. The construction was obtained by the probabilistic search technique **simulated annealing.**

3. The Projective Semi-Linear Group

The $APA_2(2, 10, 10)$ as constructed in the previous section is contained in the projective semi-linear group $P\Gamma L_2(9)$. More precisely the group $\langle\mathcal{A}\rangle$ generated by $\mathcal{A}$ is $PSL_2(9)\langle\phi\rangle$, where $PSL_2(9) \cong A_6$ is the special linear group and ϕ is the Frobenius automorphism of $\mathbb{F}_9$ over $\mathbb{F}_3$. The second author conjectures that this construction generalizes as follows:

CONJECTURE 1 *Let q be an odd prime-power. Then there is a subset $\mathcal{A} \subset P\Gamma L_2(q^2)$ such that $\mathcal{A}$ is an $(Z_{(q^2+1)/2}, E_{q^2})$-admissible*

$$APA_{q-1}(2, q^2+1, q^2+1).$$

Here $Z_{(q^2+1)/2}$ and E_{q^2} denote the cyclic respectively elementary abelian subgroup of $PSL_2(q^2)$ of the corresponding orders.

The conjecture has been verified for $q \leq 9$.

PROPOSITION 1 *There exist*

- $APA_4(2, 26, 26) \subset P\Gamma L_2(25)$
- $APA_6(2, 50, 50) \subset P\Gamma L_2(49)$
- $APA_8(2, 82, 82) \subset P\Gamma L_2(81)$

We mention some more $APA_\mu(t, n, n)$, where μ is small without being optimal: The unitary group $U_3(5) = PSU_3(5^2)$ is an $APA_{16}(2, 126, 126)$, the smallest Ree group ${}^2G_2(3) \cong P\Gamma L_2(8)$ is an $APA_4(2, 28, 28)$, whereas ${}^2G_2(27)$ is an $APA_{52}(2, 19684, 19684)$. The smallest Suzuki group ${}^2B_2(8)$ is an $APA_{16}(2, 65, 65)$ and ${}^2B_2(32)$ is an $APA_{62}(2, 1025, 1025)$. Further $PSL_2(8)$ is an $APA_4(4, 9, 9)$ and $P\Gamma L_2(32)$ is an $APA_4(4, 33, 33)$.

4. Some Authentication Perpendicular Arrays

Let $\mathcal{A}$ be an $APA_\lambda(2, k, v)$. The transitive kernel $C_0(\mathcal{A})$ was defined in [2] as the set of columns c which satisfy that for every column $c' \neq c$ the restriction $\mathcal{A}_{\{c,c'\}}$ of $\mathcal{A}$ to columns c and c' is an ordered design $OD_{\lambda/2}(2, 2, v)$. It was proved that for $c \in C_0(\mathcal{A})$ the restriction of $\mathcal{A}$ to $C - \{c\}$ is an $APA_\lambda(2, k-1, v)$. We improve on [2], Proposition 3 and Corollary 15:

PROPOSITION 2 *Let $\mathcal{A}$ be an $APA_2(2, n, n)$, which is $(G, 1)$-admissible, where the group G of order $n-1$ fixes one column c and transitively permutes the remaining columns. Then $c \in C_0(\mathcal{A})$.*

Proof. It is easily seen that for every column $c' \neq c$ and every pair a, b of entries there is a row of $\mathcal{A}$ having a in column c and b in column c'. As the number of rows of $\mathcal{A}$ is $n(n-1)$, it follows that $\mathcal{A}_{\{c,c'\}}$ is an $OD_1(2, 2, n)$. ■

Application of this to our constructions of $APA_2(2, 6, 6)$, $APA_2(2, 10, 10)$ and APA_2 $(2, 12, 12)$ yields Corollary 2.

References

1. J. Bierbrauer, The uniformly 3-homogeneous subsets of $PGL_2(q)$, *Journal of algebraic combinatorics*, Vol. 4 (1995) pp. 99–102.
2. J. Bierbrauer and Y. Edel, Theory of perpendicular arrays, *Journal of Combinatorial Designs*, Vol. 6 (1994) pp. 375–406.
3. J. Bierbrauer and Y.Edel, Halving $PSL_2(q)$, to appear in *Journal of Geometry.*
4. J. Bierbrauer and T. v. Tran, Halving $PGL_2(2^f)$, f odd: a Series of Cryptocodes, *Designs, Codes and Cryptography*, Vol. 1 (1991) pp. 141–148.
5. J. Bierbrauer, T. v. Tran, Some highly symmetric Authentication Perpendicular Arrays, *Designs, Codes and Cryptography*, Vol. 1 (1992) pp. 307–319.
6. E. S. Kramer, D. L. Kreher, R. Rees, and D. R. Stinson, On perpendicular arrays with $t \geq 3$, *Ars Combinatoria*, Vol. 28 (1989) pp. 215–223.
7. C. R. Rao, Combinatorial Arrangements analogous to Orthogonal Arrays, *Sankhya A*, Vol. 23 (1961) pp. 283–286.
8. D. R. Stinson, The Combinatorics of Authentication and Secrecy Codes, *Journal of Cryptology*, Vol. 2 (1990) pp. 23–49.
9. D. R. Stinson and L. Teirlinck, A Construction for Authentication/Secrecy Codes from 3-homogeneous Permutation Groups, *European Journal of Combinatorics*, Vol. 11 (1990) pp. 73–79.

Designs, Codes and Cryptography, 9, 39–49 (1996)
© 1996 Kluwer Academic Publishers, Boston. Manufactured in The Netherlands.

Forcing Linearity on Greedy Codes

JEFFREY T. BONN
Department of Mathematics, Southern Illinois University, Carbondale, IL 62901 U.S.A.

Communicated by: V. Tonchev

Received October 26, 1994; Accepted August 15, 1995

Abstract. Described in this paper are two different methods of forcing greedy codes to be linear over arbitrary finite fields. Both methods are generalizations of the binary B-greedy codes as well as the triangular greedy codes over arbitrary fields. One method generalizes to arbitrary orderings over arbitrary finite fields while the other method generalizes the B-greedy codes to linear codes over arbitrary finite fields. Examples of the first method are computed for triangular greedy codes. These examples give codes similar to triangular codes from B-orderings. Both methods are shown to be substantially different.

1. Introduction

A greedy code, C, is a q-ary (n, m, d) code generated by applying to some ordering of A_q^n (the length n strings of elements from $\{0, 1, \ldots, q-1\}$) a greedy algorithm that retains, for the code, the minimum distance d. The greedy algorithm picks the first word in the ordering to be in C. Then it continues through the ordering, checking to see what the minimum distance is between a word and the words in C. Once a word is found whose minimum distance is at least d from everything in C, it is included in the code C. So it recursively proceeds through the ordered list, adding valid words into the code. This guarantees that the minimum distance $d(C) \geq d$.

A B-ordering of $\mathbf{F}_2^n$ is an ordering arising from an ordered basis, B, of $\mathbf{F}_2^n$ by recursively adding all previous words to the next basis word. The first word in the ordering is always the zero word. See Example 1 for a B-ordering and the greedy code that arises from it, referred to as a B-greedy code.

Example 1. This example uses $d = 2$ and $B = \{1000, 0100, 0010, 0001\}$. The basis vectors are indicated by an arrow ($\rightarrow$) and the code words by a diamond ($\diamond$). The B-greedy code arising here is a binary [4,3,2] code.

0000 $\diamond$		$\rightarrow$	0001
$\rightarrow$	1000		1001 $\diamond$
$\rightarrow$	0100		0101 $\diamond$
	1100 $\diamond$		1101
$\rightarrow$	0010		0011 $\diamond$
	1010 $\diamond$		1011
	0110 $\diamond$		0111
	1110		1111 $\diamond$

It has been shown in different ways that B-greedy codes are linear ([1], [2], [4], [5], [7]). Brualdi and Pless [2] showed it by constructing a linear map via so called g-values computed in a greedy algorithm. Directly obtained from the g-values was a parity check matrix.

Of particular interest in the paper of Brualdi and Pless is the introduction of triangular greedy codes. These are B-greedy codes with triangular bases. Using a computer, they checked lower dimensional cases for 3 different classes of triangular bases and found that the codes were always at most one away from (but usually exactly the same as) the dimension of the biggest known code. They produced some bounds for dimensions of triangular greedy codes, but these were not nearly as close as the computer data suggested they could be.

Brualdi and Pless [2] also gave an algorithm that directly generates the parity check matrix for triangular greedy codes, given a triangular basis. This algorithm is valid over arbitrary finite fields. Monroe [6] has used this algorithm to construct record breaking ternary and quaternary codes. Since there is no theory explaining why such good results are being achieved, more study is certainly warranted. In this paper, two substantially different generalizations of B-greedy codes are given which are linear. Examples of one generalization are given. These examples provide more evidence that it is the triangular nature of the orderings, rather than the B-ordering structure, that make the triangular greedy codes so big.

2. Linearity Forced on Greedy Codes

There are examples of greedy codes which are not linear (See Example 2). When a new codeword is added to the code, linearity can be forced by adding all multiples of the new codeword to the previous codewords. Since the zero word must also be in a linear code, it is always placed first in the ordering so inclusion is guaranteed. Codes generated by this algorithm that forces linearity will be called *forced greedy codes*. Basically, this is greedily picking the basis for a code rather than greedily picking the whole code. See Example 2 for an example of a forced greedy code.

Forcing linearity will produce linear codes for any ordering (that begins with the zero word) over arbitrary finite fields. However, B-orderings do not produce linear codes over arbitrary finite fields as Example 3 shows. For ternary codes there are two logical candidates for B-orderings, the one related to the lexicographic order being more natural than the other. Example 3 contains both. Note that Examples 2 and 3 give cases where the forced greedy code and the greedy code arising from the same ordering can be either smaller (Example 2) or bigger (Example 3).

Example 2. This example is over $\mathbf{F}_2$ with $d = 2$ which compares the forced greedy algorithm, denoted by a heart ($\heartsuit$), and the greedy algorithm, denoted by a diamond ($\diamondsuit$).

0000	◇	♡	0001		
1000			1001		
0100			0101		♡
1110	◇	♡	1101	◇	
0010			0011		
1011	◇	♡	010		
0111	◇		0110		
1100			1111		

Example 3. Two pairs of examples of nonlinear ternary greedy codes and forced greedy codes with $d = 2$ and $B = \{100, 010, 001\}$. Notation is combined from Examples 1 and 2.

	000	◇	♡			
→	100			211		♡
	200			021		♡
→	010			121		
	110	◇	♡	221		
	210			002		
	020			102		
	120			202	◇	♡
	220	◇	♡	012		♡
→	001			112		
	101	◇	♡	212		
	201			022	◇	
	011	◇		122		♡
	111			222		

	000	◇	♡			
→	100			202	◇	♡
	200			011	◇	
→	010			012		♡
	020			021		♡
	110	◇	♡	022	◇	
	120			111		
	210			112		
	220	◇	♡	121		
→	001			122		
	002			211		♡
	101	◇	♡	212		
	102			221		
	201			222		

With forced linearity we are adding words to the code which have not been checked for the minimum distance condition. The next proposition assures that the condition is still satisfied.

PROPOSITION 1 *Let C be a q-ary linear code with* $\mathrm{d}(C) \geq d$, *and* $\mathbf{x}$ *such that* $\mathrm{d}(\mathbf{x}, C) \geq d$. *Then* $d(\{i\mathbf{x}+\mathbf{y} : \mathbf{y} \in C, i \in \mathbf{F}_q\}) \geq d$.

Proof. Let $\mathbf{y},\mathbf{z} \in C$ and $i, j \in \mathbf{F}_q$. Then

$$
\begin{aligned}
\mathrm{d}(\{i\mathbf{x}+\mathbf{y}, j\mathbf{x}+\mathbf{z}\}) &= \mathrm{w}([i\mathbf{x}+\mathbf{y}]-[j\mathbf{x}+\mathbf{z}]) \\
&= \mathrm{w}([i-j]\mathbf{x}-[\mathbf{z}-\mathbf{y}]) \\
&= \begin{cases} \mathrm{w}(\mathbf{y}-\mathbf{z}), & \text{if } i=j \\ \mathrm{w}(\mathbf{x}-[i-j]^{-1}[\mathbf{z}-\mathbf{y}]), & \text{if } i \neq j \end{cases} \\
&= \begin{cases} \mathrm{d}(\mathbf{y},\mathbf{z}), & \text{if } i=j \\ \mathrm{d}(\mathbf{x},[i-j]^{-1}[\mathbf{z}-\mathbf{y}]), & \text{if } i \neq j \end{cases} \\
&\geq d, \text{ as } [i-j]^{-1}[\mathbf{z}-\mathbf{y}] \in C \text{ (because C is linear),} \\
&\quad \text{and } \mathrm{d}(\mathbf{x}, C) \geq d.
\end{aligned}
$$

□

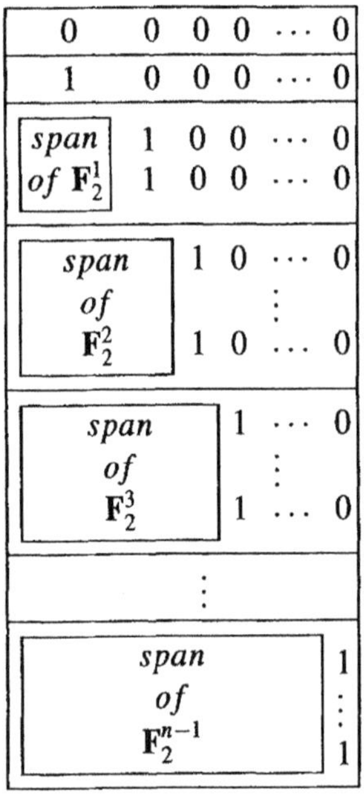

Figure 1.

3. Codes From Generalized Triangular Orderings of $\mathbf{F}^n_\Theta$

Brualdi and Pless in [2] produced the biggest codes nearly all the time with triangular greedy codes. Figure 1 is a generalization of the triangular orderings of $\mathbf{F}_2^n$ they used, i.e., B-orderings with a triangular basis. All spans referred to are arbitrary orderings. Each dimension is separated by a horizontal line. Note that these orderings can be generalized to $\mathbf{F}_q^n$ by inserting, for each dimension, the suitable multiples in any order. Therefore, forced triangular greedy codes (forced greedy codes from a generalized triangular ordering) are a generalization of the triangular greedy codes over arbitrary finite fields.

Some examples of the binary forced triangular greedy codes have been computed to compare with the triangular greedy codes from Brualdi and Pless. All examples use generating patterns for the various spans in the generalized triangular ordering. If $p(x) = x^n + \sum_{i=0}^{n-1} b_i x^i$ is a primitive polynomial over $\mathbf{F}_q$ and $\{\mathbf{v}_1, \ldots, \mathbf{v}_n\}$ is a basis of $\mathbf{F}_q^n$, then an ordering of $\mathbf{F}_q^n$ can be generated recursively for $k > n$ by

$$\mathbf{v}_k = \sum_{i=0}^{n-1} b_i \mathbf{v}_{k-n+i}$$

Fitzgerald and Yucas have shown in [3] that these generating patterns $(b_0, \ldots, b_{n-1})$ do in fact generate the entire non-zero span. Forced greedy codes have been computed from these generating pattern orderings alone and no relationship seems to exist between the generating patterns and the size of the codes. However, the spans are easy to generate since only n words need to be in memory to generate the next word in the span. These spans were used in the triangular ordering because they are more easily generated than a random ordering but seem to, by themselves, have no effect on the forced greedy code.

Table 1 gives the dimension for a single example of a forced greedy code corresponding to a length n and a minimum distance d. For some cases, a second example is denoted in parentheses. The first examples were limited by the data structure which held the code; therefore, these examples stop with dimension 12. The second examples were not so limited but limited rather by computing time, thus the sparseness of these second examples. It is easy to see that forced triangular greedy codes of even minimum distance have codewords of even weight by the same proof of Brualdi and Pless for triangular greedy codes. Thus forced greedy codes with odd minimum distance are related to even ones by adding an overall parity check bit to the code and the ordering. Table 1 does not reflect this relationship because the orderings used were not related by adding a parity check bit.

Data from this table has been compared to the table of Brualdi and Pless which computes the dimensions of lexicodes, complementary greedy codes, and Gray greedy codes. The dimensions in this table are identical to the Gray greedy codes with the exception of the second examples with $(n, d) = (17, 6), (18, 6)$. These exceptions have dimensions equal to the complementary greedy codes. Therefore, these examples of forced triangular greedy codes are within one dimension of the best known codes. Although computer resources limited the number of trials, this is more evidence that the big sizes for triangular greedy codes are due to the triangular nature of the ordering alone rather than the B-ordering structure in combination with being triangular. Note that these forced triangular greedy codes are generalizations of triangular greedy codes over arbitrary finite fields, which Monroe [6] has used to find record breaking non-binary linear codes.

4. Forcing Linearity on g-Values

In [2], Brualdi and Pless defined a map on $\mathbf{F}_2^n$ using what they called g-values. They showed this map to be linear with a B-greedy code as its kernel. Therefore the code was linear. The map, when viewed as a linear transformation on the span of the unit vectors, was then a parity check matrix for the code. These g-values were a natural extension of the greedy algorithm where distances are checked against only previous code words. For g-values, distances are checked against the code words in each coset as well. A formal definition of the g-values, $g(\mathbf{z}_i)$, follows.

Let d be the minimum distance in the greedy algorithm, and let $\mathbf{z}_i$ be the words in the ordering. Then define $g(\mathbf{z}_i)$ as the smallest nonnegative integer, t, such that $\mathrm{d}(\mathbf{z}_i, \mathbf{x}) \geq d$ for all vectors $\mathbf{x}$ in $\{\mathbf{z}_1, \ldots, \mathbf{z}_{i-1}\}$ such that $g(\mathbf{x}) = t$. If no such t exists, then $g(\mathbf{z}_i)$ is the smallest value not yet used.

Note that one can always obtain the columns of a parity check matrix for the code by considering in binary the g-values for the unit vectors.

$n:d$	3	4	5	6	7	8	9	10	11	12
3	1	0	0(0)	0(0)	0	0	0	0	0	0
4	1	1	0(0)	0(0)	0	0	0	0	0	0
5	2	1	1(1)	0(0)	0	0	0	0	0	0
6	3	2	1(1)	1(1)	0	0	0	0	0	0
7	4	3	1(1)	1(1)	1	0	0	0	0	0
8	4	4	2(2)	1(1)	1	1	0	0	0	0
9	5	4	2(2)	2(2)	1	1	1	0	0	0
10	6	5	3(3)	2(2)	1	1	1	1	0	0
11	7	6	4(4)	3(3)	2	1	1	1	1	0
12	8	7	4(4)	4(4)	2	2	1	1	1	1
13	9	8	5(5)	4(4)	3	2	1	1	1	1
14	10	9	6(6)	5(5)	4	3	2	1	1	1
15	11	10	7(7)	6(6)	5	4	2	2	1	1
16	11	11	8(8)	7(7)	5	5	2	2	1	1
17	12	11	9(9)	8(7)	6	5	3	2	2	1
18		12	9(9)	9(8)	7	6	3	3	2	2
19			10(10)	9(9)	8	7	4	3	2	2
20			11(11)	10(10)	9	8	5	4	3	2
21			12(12)	11(11)	10	9	5	5	3	3
22				12(12)	11	10	6	5	4	3
23				(13)	12	11	7	6	5	4
24				(13)		12	7	7	5	5
25				(14)			8	7	6	5
26				(15)			9	8	7	6
27				(16)			9	9	7	7
28				(17)			10	9	8	7
29							11	10	8	8
30							12	11	9	8

Table 1.

Example 4. An example of g-values for a B-greedy code with $d = 3$ and $B = \{1000, 0100, 0010, 0001\}$. Notation is the same as for Example 1 with g-values on the right.

	0000 ◇	0	→	0001	4
→	1000	1		1001	5
→	0100	2		0101	6
	1100	3		1101	7
→	0010	3		0011	7
	1010	2		1011	6
	0110	1		0111	5
	1110 ◇	0		1111	4

The parity check matrix arising here is

$$\begin{bmatrix} 1 & 0 & 1 & 0 \\ 0 & 1 & 1 & 0 \\ 0 & 0 & 0 & 1 \end{bmatrix}.$$

It is also possible to compute g-values and a parity check matrix over any finite field, $\mathbf{F}_q$, when forcing linearity on an arbitrary ordering. With B-greedy codes we have linearity of g-values for free. For arbitrary orderings, the g-values must be forced into linearity. The algorithm for this starts just as before with the definition of g-values. It continues with two cases.

Case. $g(\mathbf{z}_i)$ is the smallest value not yet used. Then let

$$g(j\mathbf{z}_i + \mathbf{x}) = jg(\mathbf{z}_i) + g(\mathbf{x}),$$

for every $j \in \mathbf{F}_q^*$ and every $\mathbf{x}$ already assigned a g-value. (Note that all additions are q-ary vector additions.)

Case. $g(\mathbf{z}_i)$ is a value already assigned. Then let

$$g(\mathbf{z}_i - \mathbf{x}) = g(\mathbf{z}_i) - g(\mathbf{x}) = 0,$$

where $\mathbf{x}$ is the first word such that $g(\mathbf{z}_i) = g(\mathbf{x})$. In this case we have a new code word, $\mathbf{z}_i - \mathbf{x}$. So all multiples and sums with previous code words are added to the code, i.e., given g-value 0. In order to force linearity for the g values overall, let

$$g(\mathbf{y} + \mathbf{x}) = g(\mathbf{y}) + g(\mathbf{x}),$$

where $\mathbf{y}$ is any new code word just added in, and $\mathbf{x}$ is the first word assigned to a g-value. When either case is finished, go back to the ordering and start again with the next word in the order that does not have a g-value assigned. This process guarantees g is a linear map.

Example 5 forces linearity on g-values and is for $\mathbf{F}_3^3$ with $d = 3$, where the ordering begins with

000
100
010
001
⋮

Since most of the work in the example is done with respect to the g-values, a series of arrays is given to illustrate the process.

Example 5. Each column is a coset associated to a g-value. Each major step is separated by a horizontal line.

g-values as vectors:	01	02	10	11	12	20	21	22
code words \ g-values:	1	2	3	4	5	6	7	8
000	?	?	?	?	?	?	?	?
000	100	?	?	?	?	?	?	?
000	100	200	?	?	?	?	?	?
000	100	200	010	?	?	?	?	?
000	100	200	010	110	210	020	120	220
000	100	200	010	110	210	020	120	220
?	?	?	?	001	?	?	?	?
000	100	200	010	110	210	020	120	220
221	?	?	?	001	?	?	?	?
112	?	?	?	?	?	?	?	?
000	100	200	010	110	210	020	120	220
221	021	121	201	001	101	211	011	111
112	212	012	122	222	022	102	202	002

The parity check matrix would be

$$\begin{bmatrix} 1 & 0 & 1 \\ 0 & 1 & 1 \end{bmatrix}.$$

In the first step of the example, the first word 000 is put into the code, i.e. given g-value 0. The next step gives $g(100) = 1$. The third step is from case 1 which gives $g(200) = 2$. Similarly, the third word in the order is given g-value 3 and from case 1 the other g-values are obtained in the next step. In step six $g(001) = 4$. So from case 2, step 7 gives new codewords 221 and 112. The rest of the words have g-values forced in the last step.

As will be seen, this is as much of the order as is necessary; what determines the code is the order of the words which are linearly independent to all previous words in the order.

THEOREM 1 *For binary codes, the codes which are the kernels of the g-function arising by forcing linearity are the same as the B-greedy codes with ordered bases corresponding to the ordered set of the first linearly independent words of the ordering.*

Proof. Let $\{\mathbf{v}_1, \ldots, \mathbf{v}_n\}$ be an ordered basis of $\mathbf{F}_2^n$ and the ordered set of the first linearly independent words of an ordering. Proceed by induction on n. For $n = 1$ the codes are clearly identical. Now assume the theorem is true for $n - 1$. By the induction hypothesis, when $\mathbf{v}_n$ is the next word to receive a g-value both the B-greedy algorithm and the algorithm forcing linearity on g-values will have constructed greedy codes which are the same. Therefore the g-values assigned so far are the same. Thus $g(\mathbf{v}_n)$ will be the same also. Since g is known to be linear for B-orderings, forcing linearity will construct the same g. Thus the two codes are the same. ∎

In lieu of this theorem, these codes will be called *generalized B-greedy codes*. The linearly independent words will be referred to as an *ordered basis* for the generalized B-greedy code. Notice that the ordered basis can occur anywhere in the listing, as long as they appear in order and are the first linearly independent words.

5. Triangular Greedy Codes via a Parity Check Matrix

From [2], the greedy algorithm for triangular greedy codes with $d \geq 2$ can be viewed as an algorithm for the construction of the parity check matrix which arises from g-values. The columns of the parity check matrix can be recursively found by the following algorithm.

Let $\mathbf{H}_1 = [1]$ be the initial parity check matrix for the trivial code consisting of the zero word. Suppose a parity check matrix $\mathbf{H}_i = \mathbf{h}_1 \cdots \mathbf{h}_i]$ has been constructed for a triangular B-ordering of $\mathbf{F}_2^i$, where the columns correspond to g-values of the unit vectors. Consider another basis vector, $(x_1, \ldots, x_i, x_{i+1} = 1)$, for $\mathbf{F}_2^{i+1}$ to extend the triangular basis for the B-ordering of $\mathbf{F}_2^i$. Let $t > 0$ be the smallest integer such that when viewed as a column in binary $\mathbf{h}_{i+1} = x_1\mathbf{h}_1 + \cdots + x_i\mathbf{h}_i + x_{i+1}\mathbf{t}$ is not the sum of fewer than $d - 1$ columns of $\mathbf{H}_i$. Now let $\mathbf{H}_{i+1} = [\mathbf{h}_1 \cdots \mathbf{h}_i\ \mathbf{h}_{i+1}]$.

It has been shown by Brualdi and Pless in [2] that the t in this algorithm is the g-value for the basis vectors of the triangular greedy code. Therefore this algorithm does construct the parity check matrix of the corresponding triangular greedy code. This algorithm generalizes to arbitrary finite fields where $x_{i+1} \neq 0$. Monroe in [6] used this algorithm to construct record breaking non-binary codes. The following theorem shows that the generalized B-greedy codes are generalizations of the triangular greedy codes constructed via the parity check matrix algorithm over any finite field to an arbitrary basis. This lends even more credence to the terminology generalized B-greedy codes.

THEOREM 2 *Given a triangular ordered basis of* $\mathbf{F}_q^n$, *the generalized B-greedy algorithm constructs the same code as the parity check matrix algorithm.*

Proof. Let $\{\mathbf{v}_1, \ldots, \mathbf{v}_n\}$ be an ordered basis of $\mathbf{F}_q^n$. Just as t is the g-value for $\mathbf{v}_i$ in the parity check matrix algorithm over $\mathbf{F}_2^n$, t can be viewed as a g-value for $\mathbf{v}_i$ over $\mathbf{F}_q^n$ identical to the one found by the generalized B-greedy algorithm. The same induction can be done as in Theorem 2 to complete the proof. ■

6. Two Greedy Algorithms that Force Linearity

Note that a generalized B-greedy code may be different than a forced greedy code from the same ordering. Consider the ordering used in Example 5. For Example 6, let it continue as:

000
100
010
001
111
⋮

Let Example 6 be the forced greedy code on this ordering with $d = 2$. Now 111 will be the second codeword and forcing linearity on the code will then put 222 in the code. So the forced greedy code is $\{000, 111, 222\}$, whereas the generalized B-greedy code of Example 5 on the same ordering is $\{000, 112, 221\}$.

This would suggest that these two greedy algorithms may produce equivalent codes. However, this is not the case; for example, consider Example 7. This is a binary $[3, 2, 2]$ code, but the forced greedy code on the same ordering would be $\{000, 111\}$: a binary $[3, 1, 2]$ code. Here the generalized B-greedy code is bigger than the forced greedy code arising from the same ordering.

Example 7. The ordering for this example starts the same as for Example 6 but is an ordering for $\mathbf{F}_2^3$. Also, $d = 2$ again. The generalized B-greedy algorithm gives:

g-values as vectors:	01
code words \ g-values:	1
000	?
000	100
000	100
?	010
000	100
110	010

g-values as vectors:	01
code words \ g-values:	1
000	100
110	010
?	001
000	100
110	010
101	001
011	111

It is possible for the forced greedy code to be bigger than a generalized B-greedy code. To see an example, consider the forced greedy codes produced in Table 1: in particular, the examples where $n = 23, 24$ and $d = 9, 10$ respectively. Note that these forced greedy codes are bigger than the corresponding lexicodes from the table of Brualdi and Pless. When the forced greedy codes were computed, generating patterns were used to get the arbitrary spans of the triangular order. Note that generating patterns only give nonzero words; but that will not change the code because the zero word will only contribute a word of weight one which will never be picked for a code with $d \geq 2$. This observation can be used to place the zero word first in every arbitrary span. Thus the lexicographic basis will be the set of linearly independent words. The generalized B-greedy algorithm applied to these modified orderings will produce lexicodes. This modification will not change the forced greedy code. Therefore, this modified ordering produces a forced greedy code that is bigger than a generalized B-greedy code.

Acknowledgements

The author would like to thank J. Yucas for many helpful comments and suggestions as well as the referees for their valuable comments and suggestions.

References

1. R. A. Brualdi and V. Pless, Greedy codes, *JCT(A)*, Vol. 64 (1993) pp. 10–30.
2. J. H. Conway and N. J. A. Sloane, Lexicographic codes: Error correcting codes from game theory, *IEEE Trans. Inform. Theory*, Vol. IT-32 (1986) pp. 337–348.
3. R. Fitzgerald and J. Yucas, On generating linear spans over GF(p), *Congressus Numerantium*, Vol. 69 (1989) pp. 55–60.
4. D. Fon-Der-Flaass, A Note on greedy codes, to appear.
5. V. I. Levenstein, A Class of systematic codes, *Dokl. Akad. Nauk*, Vol. 1 (1960) pp. 368–371.
6. L. Monroe, Binary greedy codes, to appear in *Congressus Numerantium*, pp. 100–104.
7. A. J. Van Zanten, Lexicodes over fields of characteristic 2, to appear.

Design, Codes and Cryptography, 9, 51–59 (1996)
© 1996 Kluwer Academic Publishers, Boston. Manufactured in The Netherlands.

Oval Designs in Desarguesian Projective Planes

LAUREL L. CARPENTER llcarpe@prism.clemson.edu
Clemson University, Clemson SC 29634

Communicated by V. Tonchev
Received October 11, 1994; Accepted May 12, 1995

Abstract. Given any projective plane Π of even order q containing a hyperoval $\mathcal{O}$, a Steiner 2-$(\binom{q}{2}, \frac{q}{2}, 1)$ design can be constructed. The 2-rank of this design is bounded above by $\mathrm{rank}_2(\Pi) - q - 1$. Using a result of Blokhuis and Moorhouse [3], we show that this bound is met when Π is desarguesian and $\mathcal{O}$ is regular. We also show that the block graph of the Steiner 2-design in this case produces a Hadamard design which is such that the binary code of the associated 3-design contains a copy of the first-order Reed-Muller code of length 2^{2m}, where $q = 2^m$.

Keywords: hyperoval, Hadamard, oval design, Reed-Muller

1. Introduction

Two classes of designs that can be constructed from a hyperoval $\mathcal{O}$ in a projective plane Π of even order q are the oval design $\mathcal{W}(\Pi, \mathcal{O})$, with parameters 2-$(\binom{q}{2}, \frac{q}{2}, 1)$, and the Hadamard 3-design $\mathcal{H}(\Pi, \mathcal{O})$, with parameters 3-$(q^2, \frac{q^2}{2}, \frac{q^2}{4} - 1)$. In Theorem 3.3 it is shown that, in the case where Π is the desarguesian projective plane of order $q = 2^m$ and $\mathcal{O}$ is a regular hyperoval, the 2-rank of $\mathcal{W}(\Pi, \mathcal{O})$ is $3^m - 2^m$. Computational results for the 2-ranks of the oval designs for *all* hyperovals in the desarguesian projective planes of order $q = 2^m, m \leq 5$, as well as for the four known hyperovals in the plane of order 64, support the conjecture that $\mathrm{rank}_2(\mathcal{W}(\Pi, \mathcal{O})) = 3^m - 2^m$ for *any* hyperoval $\mathcal{O}$ in the desarguesian projective plane Π of order $q = 2^m$.

Another question that can be asked for designs coming from a hyperoval in a projective plane concerns the existence of the first-order Reed-Muller code $\mathcal{R}(1, 2m)$ as a subcode of the binary code of $\mathcal{H}(\Pi, \mathcal{O})$. It is shown in Section 4 that, in the case where $\mathcal{O}$ is a regular hyperoval and Π is the desarguesian projective plane of order $q = 2^m$, there does indeed exist a copy of $\mathcal{R}(1, 2m)$ inside the binary code of $\mathcal{H}(\Pi, \mathcal{O})$.

If the 2-rank of $\mathcal{W}(\Pi, \mathcal{O})$ and the dimension of $\mathrm{Hull}_2(\mathcal{W}(\Pi, \mathcal{O}))$ are both known, then the 2-rank of the Hadamard 3-design $\mathcal{H}(\Pi, \mathcal{O})$ can be computed as

$$\mathrm{rank}_2(\mathcal{H}(\Pi, \mathcal{O})) = \mathrm{rank}_2(\mathcal{W}(\Pi, \mathcal{O})) - \dim(\mathrm{Hull}_2(\mathcal{W}(\Pi, \mathcal{O}))) + 1.$$

Computational results for the dimensions of the binary code, the orthogonal code, and the hull of the oval design for each hyperoval in the desarguesian planes of order $q = 2^m, m \leq 5$, as well the 2-ranks of the Hadamard 3-designs for those same hyperovals are given in Section 5.

Definitions and terminology are given in Section 2, and the results on oval designs are in Section 3.

2. Background and Terminology

An incidence structure of points $\mathcal{P}$ and blocks $\mathcal{B}$ is a $t - (v, k, \lambda)$ design if $|\mathcal{P}| = v$, each block B in $\mathcal{B}$ is incident with exactly k points of $\mathcal{P}$, and each subset of t points is incident with exactly λ blocks. Given a design of blocks and points, an s-arc is a set of points which meets any block of the design at most twice. Blocks which meet the arc in 2, 1, and 0 points are called secants, tangents, and exterior lines respectively.

A projective plane Π of order q is a symmetric $2 - (q^2 + q + 1, q + 1, 1)$ design. It is well known that the size s of an arc in Π is at most $q + 1$ if q is odd and at most $q + 2$ if q is even. Each point of a $(q + 1)$-arc has a unique tangent; when q is even these tangents are concurrent at a point referred to as a knot (or nucleus). Any $(q + 1)$-arc in an even-ordered projective plane can be extended uniquely to form a hyperoval (a $(q + 2)$-arc) by adding the knot to the points of the $(q + 1)$-arc.

Bose and Shrikhande [4] first defined the design formed by the dual of the complement of a hyperoval $\mathcal{O}$ in a projective plane Π. These designs are known as oval designs and are denoted by $\mathcal{W}(\Pi, \mathcal{O})$.

DEFINITION 2.1 *The oval design* $\mathcal{W}(\Pi, \mathcal{O})$ *related to a hyperoval* $\mathcal{O}$ *in a projective plane* Π *is the design whose points are the exterior lines of* $\mathcal{O}$ *in* Π, *whose blocks are the points in* Π *which are not in* $\mathcal{O}$, *and whose incidence is inherited from the ambient plane.*

The proof that this is indeed a design with parameters 2-$(\binom{q}{2}, \frac{q}{2}, 1)$, where q is the order of the plane, can be found in [1, Chapter 7].

Another set of designs which arise from a hyperoval in a projective plane are Hadamard designs:

DEFINITION 2.2 *Let* Π *be a projective plane of even order* q *in which there exists a hyperoval* $\mathcal{O}$. *Define* $\mathcal{P}$ *to be the set of points in* Π *which are not on the hyperoval* $\mathcal{O}$. *Then for each* $P \in \mathcal{P}$ *define*

$$B_P = \{Q \in \mathcal{P} \mid \textit{the line } QP \textit{ is a secant of } \mathcal{O} \textit{ in } \Pi\} \cup \{P\}.$$

The points $\mathcal{P}$ *and blocks* $\mathcal{B} = \{B_P \mid P \in \mathcal{P}\}$ *form a*

$$2 - (q^2 - 1, \frac{q^2}{2} - 1, \frac{q^2}{4} - 1)$$

symmetric Hadamard design which will be denoted by $\mathcal{H}(\Pi, \mathcal{O})^*$.

This design is the complement of the design of the block graph of the oval design $\mathcal{W}(\Pi, \mathcal{O})$. For further details on this construction see Masschetti [8]. An alternate construction which gives rise to the complementary Hadamard 2-design $\overline{\mathcal{H}}(\Pi, \mathcal{O})^*$ is given in Assmus and Key [1] and is equivalent to forming the block graph of the Steiner 2-design. The Hadamard 2-design $\mathcal{H}(\Pi, \mathcal{O})^*$ can be extended uniquely to a Hadamard 3-design $\mathcal{H}(\Pi, \mathcal{O})$.

For any field F, $F^{\mathcal{P}}$ is the vector space of functions from $\mathcal{P}$ to F with basis given by the characteristic functions of the singleton subsets of $\mathcal{P}$. If $\mathcal{D} = (\mathcal{P}, \mathcal{B})$ is an incidence structure, the **code** $C_F(\mathcal{D})$ of $\mathcal{D}$ over F is the subspace of $F^{\mathcal{P}}$ spanned by the characteristic functions of the blocks of $\mathcal{D}$. If $F = F_p$ the code $C_F(\mathcal{D})$ is also denoted by $C_p(\mathcal{D})$ and its dimension is referred to as the **p-rank** of $\mathcal{D}$. As usual the orthogonal code, $C^\perp$, is taken with respect to the standard inner product in F^v, i.e., for $u, w \in F^v$, $(u, w) = \sum_{x \in \mathcal{P}} u(x)w(x)$. Thus

$$C^\perp = \{u \mid u \in F^v \text{ and } (u, w) = 0 \text{ for all } w \in C\}.$$

Further, for $\mathcal{D} = (\mathcal{P}, \mathcal{B})$, and F any field, define the hull of $\mathcal{D}$ at F:

$$\text{Hull}_F(\mathcal{D}) = C_F(\mathcal{D}) \cap C_F(\mathcal{D})^\perp.$$

Two codes are **isomorphic** if each can be obtained from the other by a permutation of the coordinate positions.

The following well-known facts concerning the design of points and hyperplanes of a $(2m-1)$-dimensional projective geometry, $PG_{2m-1}(2)$, over the finite field F_2 will be assumed:

RESULT 2.3 *The incidence structure of points and hyperplanes of $PG_{2m-1}(2)$ is a Hadamard $2-(2^{2m}-1, 2^{2m-1}-1, 2^{2m-2}-1)$ design, and its binary code is the punctured first-order Reed-Muller code $\mathcal{R}(1, 2m)^*$.*

The finite field of q elements will be denoted by F_q and the multiplicative group $F_q \setminus \{0\}$ will be denoted by F_q^*. Define D to be a cyclic $(q-1, \frac{q}{2}-1, \frac{q}{4}-1)$ difference set in F_q^* and let $E = D \cup \{0\}$. Denote the 2-dimensional vector space over F_q by V and its dual space by V^t. Construct a design $\mathcal{H}(D)$ using the non-zero vectors of V as points and the non-zero linear functionals of V^t as blocks and defining a point $x \in V \setminus \{(0,0)\}$ to be incident with a block $L \in V^t \setminus \{(0,0)^t\}$ if $L \cdot x \in E$. Jackson [6] showed $\mathcal{H}(D)$ to be a symmetric $2-(q^2-1, \frac{q^2}{2}-1, \frac{q^2}{4}-1)$ Hadamard design.

Recall from field theory that the absolute trace function is a mapping from F_q $(q = 2^m)$ onto F_2 defined by $tr(\alpha) = tr_{2^m/2}(\alpha) = \alpha^{2^{m-1}} + \alpha^{2^{m-2}} + \ldots + \alpha^2 + \alpha$ for any $\alpha \in F_q$. Let $C_0^* = \{\alpha \mid \alpha \in F_q^*, tr(\alpha) = 0\}$ and define $C_0^{*-1} = \{\alpha^{-1} \mid \alpha \in C_0^*\}$.

LEMMA 2.4 *The sets C_0^* and C_0^{*-1} are both cyclic $(q-1, \frac{q}{2}-1, \frac{q}{4}-1)$ difference sets.*

Proof: The finite field F_{2^m} can be viewed as a vector space V of dimension m over F_2 with basis $e, \omega, \omega^2, \ldots, \omega^{m-1}$ where ω is a primitive element of F_{2^m} and e is the identity element of F_{2^m}. The absolute trace map tr from F_{2^m} onto F_2 is a linear transformation with kernel $\ker(tr) = C_0$ of dimension 2^{m-1} over F_2. Thus, C_0^* has $2^{m-1}-1$ elements and can be viewed geometrically as a hyperplane of $PG_{m-1}(2)$. The linear transformation S defined by mapping $v \in V$ to $\omega v \in V$ is an automorphism of V which acts cyclically on the points of $PG_{m-1}(2)$. S also acts as a cyclic automorphism on the hyperplanes of $PG_{m-1}(2)$.

It now remains to be shown that for each $\eta \neq e$ in $F_{2^m}^*$ there exists exactly $\lambda = 2^{m-2}-1$ pairs (c_1, c_2) in $C_0^* \times C_0^*$ such that $c_1 c_2^{-1} = \eta$. View η and e as points in $PG_{m-1}(2)$. Two points define a unique line and each line is in λ hyperplanes. Thus η and e are together in hyperplanes $h_1, h_2, \ldots, h_\lambda$. Because of the action of S on V^t, for each $i = 1, \ldots, \lambda$ there exists a unique ω^i in $F_{2^m}^*$ such that $h_i = C_0^*\omega^i$. Therefore, $\eta\omega^{-i}$ and ω^{-i} are in C_0^* for all $i = 1, \ldots, \lambda$. Further $(\eta\omega^{-i})(\omega^{-i})^{-1} = \eta$, so there are at least $2^{m-1} - 1$ solutions in $C_0^* \times C_0^*$ to $c_1c_2^{-1} = \eta$. There are $2^m - 1$ choices for η so the solutions already enumerated account for $(2^m - 1)(2^{m-2} - 1)$ pairs (c_1, c_2) in $C_0^* \times C_0^*$. The $2^{m-1} - 1$ pairs remaining are those in which $c_1 = c_2$ and thus will not add to the possible solutions since $\eta \neq e$. Hence, C_0^* is a cyclic difference set.

Since C_0^* is a difference set, $c_1c_2^{-1} = \eta$ has exactly λ solutions (c_1, c_2) in $C_0^* \times C_0^*$ for every element $\eta \neq e \in F_q^*$. Therefore, for every non-identity element $\eta^{-1} \in F_q^*$ the equation $(c_1^{-1})(c_2^{-1})^{-1} = \eta^{-1}$ has exactly λ solutions (c_1^{-1}, c_2^{-1}) in $C_0^{*-1} \times C_0^{*-1}$. Hence, C_0^{*-1} is also a difference set with the same parameters as C_0^*. □

In [6] it was shown that

$$\mathcal{H}(C_0^*) \cong PG_{2m-1}(2)$$

and

$$\mathcal{H}(C_0^{*-1}) \cong \mathcal{H}(PG_2(2^m), \mathcal{R})^*$$

where $\mathcal{R}$ is a regular hyperoval. Incidence matrices for these designs may be labeled with the non-zero vectors of V by representing the jth column as x_j and the ith row as y_i where the x_j's and y_i's range over the vectors in $V \setminus \{(0,0)\}$.

An incidence matrix for $\mathcal{H}(C_0^*)$ is given by $G = [g_{ij}]$ where

$$\begin{aligned} g_{ij} &= \begin{cases} 1 \text{ if } x_jy_i^T \in C_0 \\ 0 \text{ if } x_jy_i^T \notin C_0 \end{cases} \\ &= \begin{cases} 1 \text{ if } tr(x_jy_i^T) = 0 \\ 0 \text{ if } tr(x_jy_i^T) = 1 \end{cases}. \end{aligned}$$

Likewise, an incidence matrix for $\mathcal{H}(C_0^{*-1})$ is given by $M = [m_{ij}]$ where

$$\begin{aligned} m_{ij} &= \begin{cases} 1 \text{ if } x_jy_i^T \in (C_0^*)^{-1} \cup \{0\} \\ 0 \text{ if } x_jy_i^T \notin (C_0^*)^{-1} \cup \{0\} \end{cases} \\ &= \begin{cases} 1 \text{ if } tr(x_jy_i^T)^{q-2} = 0 \\ 0 \text{ if } tr(x_jy_i^T)^{q-2} = 1 \end{cases}. \end{aligned}$$

3. Oval Designs

In 1989 Mackenzie [7] proved that for any oval design $\mathcal{W}(\Pi, \mathcal{O})$ from a plane of order $q = 2^m$, $\text{rank}_2(\mathcal{W}(\Pi, \mathcal{O})) \leq 3^m - 2^m$. At that time it was conjectured that the 2-rank of an oval design, $\mathcal{W}(\Pi, \mathcal{O})$, from any projective plane Π of order $q = 2^m$ is given by $\text{rank}_2(\mathcal{W}(\Pi, \mathcal{O})) = \text{rank}_2(\Pi) - 2^m - 1$. However, Mackenzie disproved the conjecture by finding a counter-example in one of the translation planes of order 16. A weaker form of the conjecture may still prove to be true.

CONJECTURE 3.1 *The 2-rank of an oval design $\mathcal{W}(\Pi, \mathcal{O})$ from the desarguesian projective plane of order $q = 2^m$ is given by* $\mathrm{rank}_2(\mathcal{W}(\Pi, \mathcal{O})) = 3^m - 2^m$.

For regular hyperovals in desarguesian projective planes the conjecture can be proven as a corollary to the following result by Blokhuis and Moorhouse [3]:

RESULT 3.2 *For a regular hyperoval in the desarguesian projective plane of order $q = 2^m$, the 2-rank of the incidence matrix of nonsecants is* $3^m + 1$.

In order to show the the truth of Conjecture 3.1 for regular hyperovals in desarguesian projective planes, let $\mathcal{O}$ represent a regular hyperoval in the desarguesian plane Π of order $q = 2^m$. Then $\mathcal{O}$ is a conic $\mathcal{Q}$ of $s = 2^m + 1$ points along with its nucleus. Let $P_1, P_2, \ldots, P_s$ denote the points of $\mathcal{Q}$ and let P_{s+1} denote the nucleus of $\mathcal{Q}$. Then $P_{s+2}, \ldots, P_v$ are the remaining points in Π. Also let $l_1, l_2, \ldots, l_t$ be the lines in Π which are secant to $\mathcal{Q}$. There are $t = \frac{s(s-1)}{2}$ of these. Let $l_{t+1}, \ldots l_{t+s}$ be the lines in Π where l_{t+i} is tangent to $\mathcal{Q}$ at P_i. Finally, let $l_{t+s+1}, \ldots, l_v$ denote the remaining lines of Π. The incidence matrix A of points and lines of Π can be partitioned as

$$\begin{bmatrix} A_1 \\ A_2 \\ A_3 \end{bmatrix} = \begin{bmatrix} A_{11} & A_{12} & A_{13} \\ A_{21} & A_{22} & A_{23} \\ A_{31} & A_{32} & A_{33} \end{bmatrix}$$

where the rows of A_1 are the secants of $\mathcal{Q}$, the rows of A_2 are the tangents of $\mathcal{Q}$, and the rows of A_3 are the lines exterior to $\mathcal{Q}$. The submatrix $\begin{bmatrix} A_2 \\ A_3 \end{bmatrix}$ is the incidence matrix of nonsecants. By Result 3.2 the 2-rank of $\begin{bmatrix} A_2 \\ A_3 \end{bmatrix}$ is $3^m + 1$. That is, the binary code of the design of points and lines in Π is spanned by the incidence vectors of the nonsecants of $\mathcal{Q}$.

The submatrix A_{21} is the $s \times s$ identity matrix where $s = 2^m + 1$. Thus the 2-rank of A_2 must be $2^m + 1$. Suppose α is in the span of A_2. Then α is either the zero vector or has nonzero entries in the first s coordinates. If the latter is the case, then α is not in the span of A_3. Similarly, if $\beta \neq 0$ is in the span of A_3 then β has all zero entries in the first $s + 1$ coordinates. Thus β could not be in the span of A_2. Therefore, $\langle A_2 \rangle \cap \langle A_3 \rangle = \{0\}$. However, $\langle A_2 \rangle \oplus \langle A_3 \rangle = \langle A \rangle$. Hence $\mathrm{rank}_2(A_3) = 3^m + 1 - \mathrm{rank}_2(A_2) = (3^m + 1) - (2^m + 1) = 3^m - 2^m$.

Let C' be the subspace of $\langle A \rangle$ spanned by the set of the lines exterior to $\mathcal{Q}$, i.e. the subspace spanned by A_3. Now $\dim(C') = \mathrm{rank}_2(A_3) = 3^m - 2^m$, and since A_{31} and A_{32} are both zero matrices, $\mathrm{rank}_2(A_3) = \mathrm{rank}_2(A_{33})$. The incidence matrix for $\mathcal{W}(\Pi, \mathcal{O})$ is A_{33}^T. Hence, $\mathrm{rank}_2(\mathcal{W}(\Pi, \mathcal{O})) = \mathrm{rank}_2(A_{33}^T) = \mathrm{rank}_2(A_{33}) = 3^m - 2^m$. Thus the following has been shown:

THEOREM 3.3 *Let $\mathcal{O}$ be a regular hyperoval in a desarguesian plane Π of order $q = 2^m$. The 2-rank of $\mathcal{W}(\Pi, \mathcal{O})$ is $3^m - 2^m$.*

4. Hadamard Designs

The following question was posed in a more general form by Assmus and Key [2]:

QUERY 4.1 *Given a projective plane of order $q = 2^m$ containing a hyperoval $\mathcal{O}$, does the binary code of the Hadamard 3-design $\mathcal{H}(\Pi, \mathcal{O})$ arising from the hyperoval contain a copy of the first-order Reed-Muller code $\mathcal{R}(1, 2m)$?*

In the special case where Π is desarguesian and the hyperoval $\mathcal{O}$ is regular, Query 4.1 can be answered in the affirmative. This can be shown by imposing an ordering on the rows and columns of the matrices G and M, as defined in Section 2, and using Jackson's results. The idea of applying an ordering to G and M is due to Norwood [9] who used a similar ordering to prove a conjecture by Assmus and Key [1, Conjecture 7.12.1] regarding the 2-rank of a Hadamard 3-design arising from a regular hyperoval in a desarguesian projective plane. Norwood proved the following:

RESULT 4.2 *If Π is the desarguesian projective plane of order $q = 2^m$ and $\mathcal{O}$ is a regular hyperoval in Π then the 2-rank of $\mathcal{H}(\Pi, \mathcal{O})$ is $2^{m-1}m + 1$.*

Let ω be an arbitrary but fixed primitive element of F_q^* and impose an ordering of the rows and columns of G by partitioning the non-zero vectors from V so that V_∞ is the set of multiples of the vector $(0, 1)$, V_0 is the set of multiples of $(1, 0)$, and for each i such that $1 \leq i \leq q - 1$, V_i is the set of multiples of $(1, \omega^i)$. Further, order the elements of each V_i so that the kth element of V_i is $v_{i_k} = \omega^k(1, \omega^i)$. Similarly order V_∞ and V_0. Define the same ordering on the rows of G except denote it by R_∞, R_0, R_i, etc. Finally, interchange the roles of 0 and 1 as entries of the matrix. Denoting the new matrix by $G^\sim$, notice that it, together with the all-one vector, will generate a binary code isomorphic to that generated by G, since the latter does contain the all-one vector.

Impose a similar ordering of the rows and columns of M with the difference that, for each i such that $1 \leq i \leq q - 1$, V_i is the set of multiples of $(1, w^{-i})$ and the kth element of V_i is $v_{i,k} = \omega^{-k}(1, w^{-i})$. Likewise re-order the elements of V_∞, V_0, R_∞, and R_0 and reverse the entries 0 and 1. This new matrix, $M^\sim$, will, together with the all-one vector, generate a code isomorphic to that generated by M.

The matrices $G^\sim$ and $M^\sim$ are $(q^2 - 1) \times (q^2 - 1)$ block matrices where the blocks are each $(q - 1) \times (q - 1)$ and are defined by the partitioning V_∞, V_0, V_i, R_∞, R_0, R_i of columns and rows. Denote these blocks in the usual manner as $G_{i,j}^\sim$ (respectively $M_{i,j}^\sim$) where $i, j \in Z_q \cup \{\infty\}$.

Let k and ℓ be aribitrary but fixed integers from 1 to $q - 1$. The $(k, \ell)^{th}$ entry in both $G_{\infty,0}^\sim$ and $G_{0,\infty}^\sim$ is $tr(0) = 0$. Thus $G_{\infty,0}^\sim$ and $G_{0,\infty}^\sim$ are both the $(q - 1) \times (q - 1)$ zero matrix. The $(k, \ell)^{th}$ entry in $G_{\infty,\infty}^\sim$ is

$$tr(\omega^\ell(0, 1)\omega^k \begin{pmatrix} 0 \\ 1 \end{pmatrix}) = tr(w^{\ell+k}).$$

For any $1 \leq i \leq q - 1$, the $(k, \ell)^{th}$ entry in $G_{\infty,i}^\sim$ is

$$tr(\omega^{\ell}(1,\omega^{i})\omega^{k}\begin{pmatrix}0\\1\end{pmatrix}) = tr(w^{\ell+i+k})$$

and the $(k,\ell)^{th}$ entry in $G^{\sim}_{0,i}$ is

$$tr(\omega^{\ell}(1,\omega^{i})\omega^{k}\begin{pmatrix}1\\0\end{pmatrix}) = tr(w^{\ell+k}).$$

Now consider $M^{\sim}$. The $(k,\ell)^{th}$ entry in both $M^{\sim}_{\infty,0}$ and $M^{\sim}_{0,\infty}$ is $tr(0) = 0$. Thus $M^{\sim}_{\infty,0}$ and $M^{\sim}_{0,\infty}$ are both the $(q-1)\times(q-1)$ zero matrix. The $(k,\ell)^{th}$ entry of $M^{\sim}_{\infty,\infty}$ is

$$tr(\omega^{-\ell}(0,1)\omega^{-k}\begin{pmatrix}0\\1\end{pmatrix})^{q-2} = tr(w^{-(\ell+k)})^{q-2} = tr(w^{\ell+k}).$$

For any $1 \le i \le q-1$, the $(k,\ell)^{th}$ entry in $M^{\sim}_{\infty,i}$ is

$$tr(\omega^{-\ell}(1,\omega^{-i})\omega^{-k}\begin{pmatrix}0\\1\end{pmatrix})^{q-2} = tr(w^{-(\ell+i+k)})^{q-2} = tr(w^{\ell+i+k})$$

and the $(k,\ell)^{th}$ entry in $M^{\sim}_{0,i}$ is

$$tr(\omega^{-\ell}(1,\omega^{-i})\omega^{-k}\begin{pmatrix}1\\0\end{pmatrix})^{q-2} = tr(w^{-(\ell+k)})^{q-2} = tr(w^{\ell+k}).$$

Hence, the following equality has been shown:

LEMMA 4.3 *Given the matrices $G^{\sim}$ and $M^{\sim}$ defined above,*

$$\begin{bmatrix} G^{\sim}_{\infty,\infty} & G^{\sim}_{\infty,0} & \cdots & G^{\sim}_{\infty,i} & \cdots \\ G^{\sim}_{0,\infty} & G^{\sim}_{0,0} & \cdots & G^{\sim}_{0,i} & \cdots \end{bmatrix} = \begin{bmatrix} M^{\sim}_{\infty,\infty} & M^{\sim}_{\infty,0} & \cdots & M^{\sim}_{\infty,i} & \cdots \\ M^{\sim}_{0,\infty} & M^{\sim}_{0,0} & \cdots & M^{\sim}_{0,i} & \cdots \end{bmatrix}.$$

Another result concerning the first $2q-2$ rows of $G^{\sim}$ follows as a corollary to Result 2.3.

LEMMA 4.4 *The matrix*

$$G' = \begin{bmatrix} 1\ldots & 1\ldots & \ldots & 1\ldots & \ldots \\ G^{\sim}_{\infty,\infty} & G^{\sim}_{\infty,0} & \cdots & G^{\sim}_{\infty,i} & \cdots \\ G^{\sim}_{0,\infty} & G^{\sim}_{0,0} & \cdots & G^{\sim}_{0,i} & \cdots \end{bmatrix}$$

generates a code equivalent to the punctured Reed-Muller code $\mathcal{R}(1,2m)^*$.

Proof: By Result 2.3, $G^{\sim}$ is a generator matrix for a code isomorphic to the punctured Reed-Muller code $R(1,2m)^*$. It thus suffices to show that G' is of the appropriate rank. Each of the submatrices $G^{\sim}_{i,j}$, $0 < i,j \le q-1$ and $G^{\sim}_{\infty,\infty}$ span copies of $\mathcal{R}(1,m)^*$ since the absolute trace describes all linear functionals from F_{2^m} onto F_2. Therefore,

$$G^{\sim}_{\infty,\cdot} = [G^{\sim}_{\infty,\infty}, G^{\sim}_{\infty,0}, \cdots, G^{\sim}_{\infty,i}, \cdots]$$

and

$$G_{0,\cdot}^{\sim} = \left[G_{0,\infty}^{\sim}, G_{0,0}^{\sim}, \cdots, G_{0,i}^{\sim}, \cdots\right]$$

both have rank m. Further, $G_{\infty,\cdot}^{\sim}$ and $G_{0,\cdot}^{\sim}$ are linearly independent of one another because of the existence of the zero matrix in the $G_{\infty,0}^{\sim}$ and $G_{0,\infty}^{\sim}$ positions. Thus G' possesses the correct rank, $2m + 1$. □

Lemma 4.4 coupled with Lemma 4.3 shows that a copy of the punctured first-order Reed-Muller code $\mathcal{R}(1, 2m)^*$ exists inside the code generated by $M^{\sim}$ along with the all one vector.

The matrix $M^{\sim}$ along with the all one vector and the matrix m span isomorphic codes. Thus, the span of M contains a copy of $\mathcal{R}(1, 2m)^*$. Extend M by adding a row and column of ones. Call this extended matrix T. Since M is the matrix associated with a Hadamard $2-(2^{2m}-1, 2^{2m-1}-1, 2^{2m-2}-1)$ design $\mathcal{H}(\Pi, \mathcal{O})^*$, T is the matrix associated with a Hadamard $3-(2^{2m}, 2^{2m-1}, 2^{2m-2}-1)$ design $\mathcal{H}(\Pi, \mathcal{O})$. The rows in T which correspond to the rows in

$$\begin{bmatrix} M_{\infty,\infty}^{\sim} & M_{\infty,0}^{\sim} & \cdots & M_{\infty,i}^{\sim} & \cdots \\ M_{0,\infty}^{\sim} & M_{0,0}^{\sim} & \cdots & M_{0,i}^{\sim} & \cdots \end{bmatrix},$$

together with $\jmath$, the all one vector of length 2^{2m}, produce an isomorphic copy of $\mathcal{R}(1, 2m)$. Therefore, Query 4.1 has an affirmative answer in this case:

THEOREM 4.5 *The binary code of the Hadamard 3-design arising from a regular hyperoval in the desarguesian projective plane of order $q = 2^m$ contains a copy of $\mathcal{R}(1, 2m)$.*

5. Computational Results

The following table lists the dimensions of certain binary codes of designs from the hyperovals in the desarguesian projective planes. The first column of the table lists the order q of the projective plane $PG_2(q)$ and the second column indicates the specific hyperoval according to the notation:

$\mathcal{R}$ is the regular hyperoval,
$\mathcal{L}$ is the Lunelli-Sce (Hall) hyperoval,
$\mathcal{T}$ is the translation hyperoval,
$\mathcal{S}$ is the Segre hyperoval,
$\mathcal{C}$ is the Cherowitzo hyperoval,
$\mathcal{K}$ is the O'Keefe-Penttila hyperoval, and
$\mathcal{P}$ is the Payne hyperoval.

The last four columns list the dimensions of $C = C_2(\mathcal{W}(\Pi, \mathcal{O}))$, the binary code of the oval design; $C^{\perp}$, the orthogonal code of the oval design; Hull $=$ Hull$_2(\mathcal{W}(\Pi, \mathcal{O}))$ and $C_2(\mathcal{H}(\Pi, \mathcal{O}))$, the binary code of the Hadamard 3-design associated with the hyperoval.

Plane	*Hyperoval*	C	$C^{\perp}$	Hull	$C_2(\mathcal{H}(\Pi, \mathcal{O}))$
4	$\mathcal{R}$	5	1	1	5
8	$\mathcal{R}$	19	9	7	13
16	$\mathcal{R}$	65	55	33	33
	$\mathcal{L}$	65	55	1	65
32	$\mathcal{R}$	211	285	131	81
	$\mathcal{T}$	211	285	121	91
	$\mathcal{S}$	211	285	31	181
	$\mathcal{C}$	211	285	7	205
	$\mathcal{K}$	211	285	5	207
	$\mathcal{P}$	211	285	3	209

In addition to the dimensions listed above, computation shows that for each of the four known hyperovals in $PG_2(64)$ the dimension of $C_2(\mathcal{W}(PG_2(64), \mathcal{O}))$ is 665. The non-regular hyperovals in $PG_2(64)$ which are known at this time are two families of Penttila-Pinneri and one family of Penttila-Royle hyperovals. For a treatment of the known hyperovals refer to [5, 10].

References

1. E. F. Assmus, Jr. and J. D. Key. *Designs and their Codes*. Cambridge University Press, 1992. Cambridge Tracts in Mathematics, Vol. 103.
2. E. F. Assmus, Jr. and J. D. Key. Hadamard Matrices and Their Designs: A Coding-Theoretic Approach. *Transactions of the American Mathematical Society*, 330:269-293,1992.
3. A. Blokhuis and G. E. Moorhouse. Some p-ranks Related to Orthogonal Spaces. *Journal of Algebraic Combinatorics*, to appear.
4. R. C. Bose and S. S. Shrikhande. Embedding the Complement of an Oval in a Projective Plane of even order. *Discrete Mathematics*, 6:305–312, 1973.
5. W. Cherowitzo. Hyperovals in Desarguesian planes of even order. *Annals of Discrete Mathematics*, 37:87-94, 1988.
6. Wen-Ai Jackson. A characterization of Hadamard designs with $SL(2, q)$ acting transitively. *Geometriae Dedicata*, 46:197–206, 1993.
7. K. Mackenzie. *Codes of Designs*. PhD thesis. University of Birmingham, 1989.
8. A. Maschietti. Hyperovals and Hadamard designs. *Journal of Geometry*, 44:107–116, 1992.
9. T. Norwood. Private communication.
10. T. Penttila and G. F. Royle. Classification of Hyperovals in $PG(2, 32)$. *Journal of Geometry*, 50:151–158, 1994.

Designs, Codes and Cryptography, 9, 61–70 (1996)
© 1996 Kluwer Academic Publishers, Boston. Manufactured in The Netherlands.

Concerning Difference Matrices

CHARLES J. COLBOURN
Department of Combinatorics and Optimization, University of Waterloo, Waterloo, Ontario, CANADA N2L 3G1

DONALD L. KREHER
Department of Mathematical Sciences, Michigan Technological University, Houghton, Michigan, U.S.A. 49931–1295

Communicated by: D. Jungnickel

Received August 15, 1994; Accepted May 4, 1995

Abstract. Several new constructions for difference matrices are given. One class of constructions uses pairwise balanced designs to develop new difference matrices over the additive group of GF(q). A second class of constructions gives difference matrices over groups whose orders are not (necessarily) prime powers.

Keywords: Difference matrices, pairwise balanced designs, orthogonal arrays.

1. Introduction

We employ definitions from design theory consistent with [2]. An *orthogonal array* $\mathrm{OA}_\lambda(k, n)$ is a k by λn^2 array A whose entries come from an n-element set X, so that for any $1 \le i_1 < i_2 \le k$, and any $\alpha, \beta \in X$, there are exactly λ columns in the set $\{j\colon A[i_1, j] = \alpha \text{ and } A[i_2, j] = \beta\}$. The orthogonal array is said to have *order n*, *degree k* and *index* λ.

A $(s, v; \lambda)$-*difference matrix* over the group $(G, *)$ of order s is a v by $s\lambda$ matrix D with entries from G such that the multiset

$$\{D[i_1, j] * D[i_2, j]^{-1}\colon j = 1, 2, \ldots s\lambda\}$$

for any pair of rows (i_1, i_2) contains every element of G exactly λ times. If the base group G is abelian, additive notation is often used. If D is a $(s, v; \lambda)$-difference matrix expressed multiplicatively over the group $G = \{g_1, \ldots, g_s\}$, then

$$[Dg_1, Dg_2, Dg_3, \ldots, Dg_s]$$

is an $\mathrm{OA}_\lambda(v, s)$.

Difference matrices have been studied primarily as a consequence of their uses in the construction of orthogonal arrays (see, for example, [2] and [6]). In fact, an orthogonal array $\mathrm{OA}_\lambda(v, s)$ that admits an automorphism group G acting regularly on the symbols is equivalent to a $(s, v; \lambda)$-difference matrix [7].

Jungnickel [7] established that a $(s, v; \lambda)$-difference matrix satisfies $v \le s\lambda$. Most previous research on difference matrices has concentrated on constructions with a "large" number of rows (close or equal to $s\lambda$). However, for many selections of s and λ, no constructions of this type are available. In these cases, often the best bound is obtained by using the observations that given a $(s_1, v_1; \lambda_1)$-difference matrix and a $(s_2, v_2; \lambda_2)$-difference matrix, one can produce a $(s_1, \min(v_1, v_2); \lambda_1 + \lambda_2)$-difference matrix if $s_1 = s_2$, and a $(s_1 s_2, \min(v_1, v_2); \lambda_1\lambda_2)$-difference matrix. These simple "addition" and "multiplication" constructions typically yield a number of rows far from the upper bound.

Hence it is of substantial interest to improve the lower bounds on the numbers of rows in difference matrices. We develop two types of constructions here, and demonstrate that they improve some of the lower bounds implied by the known constructions.

2. A PBD Construction

In this section we give a construction for $(s, v; \lambda)$-difference matrices over the additive group of GF(s).

A pairwise balanced design of order v and index λ is a pair $(X, \mathcal{B})$ where X is a v–element set of *points* and $\mathcal{B}$ is a family of (not necessarily distinct) subsets of X called *blocks* such that every pair of points is in precisely λ blocks. We denote a pairwise balanced design of order v and index λ by PBD(v, λ). Blocks of size one and two *are permitted.*

Let $(X, \mathcal{B})$ be a PBD(v, λ). A collection $\mathcal{P}$ of n disjoint blocks in $\mathcal{B}$ that contain all of the points of X is a *parallel class of width n*. A PBD(v, λ) is *resolvable* if it can be partitioned into parallel classes.

An ℓ by w PBD(v, λ) is a PBD(v, λ) whose blocks can be partitioned into ℓ parallel classes of maximum width w. For example, developing the parallel class

$$\{0\}, \{1, 2, 4\}, \{3, 5, 6\}$$

modulo 7 constructs a 7 by 3 PBD(7, 2). Developing the parallel class

$$\{0\}, \{3\}, \{5\}, \{6\}, \{1, 2, 4\}$$

modulo 7 constructs a 7 by 5 PBD(7, 1).

LEMMA 2.1 *If there is a $(s, v; \lambda)$-difference matrix, then there exists an $s\lambda$ by s PBD(v, λ).*

Proof. Let D be a$(s, v; \lambda)$-difference matrix over the group $(G, *)$. Let X be the set of rows of D; for each column j of D and element g of G, define $B_{j,g}$ by

$$B_{j,g} = \{x \in X\colon\ D[x, j] = g\}$$

Let $\mathcal{B} = \{B_{j,g}\colon\ j \text{ is a column of } D \text{ and } g \in G\}$. For any pair of rows (i_1, i_2) of D, the multiset

$$\{D[i_1, j] * D[i_2, j]^{-1}\colon\ j = 1, 2, \ldots, s\lambda\}$$

contains the identity of G exactly λ times; hence every pair of elements of X is in exactly λ of the $\{B_{j,g}\}$. Consequently, $(X, \mathcal{B})$ is a PBD(v, λ). Since $\mathcal{P}_j = \{B_{j,g}\colon\ g \in G\}$ is a parallel class, $(X, \mathcal{B})$ is $s\lambda$ by s. ■

We establish a partial but important converse to Lemma 2.1. A definition is required. Let $(X, \mathcal{B})$ be an ℓ by w PBD(v, λ) and let $\mathcal{P}_1, \mathcal{P}_2, \ldots, \mathcal{P}_\ell$ be its partition into ℓ parallel classes of maximum width w. Let G be a group of order $s \geq w$ and for each $j = 1, 2, \ldots \ell$ select an injection Φ_j: $\mathcal{P}_j \to G$. The parallel class

$$\mathcal{P}_j = \{S_{1,j}, S_{2,j}, \ldots, S_{n,j}\}$$

describes the j-th column of an v by ℓ matrix M with entries in G by setting

$$M[k, j] = \Phi_j(S_{i,j}) \text{ if } k \in S_{ij} \in \mathcal{P}_j.$$

We say that M is a *matrix over G belonging to the ℓ by w PBD(v, λ) $(X, \mathcal{B})$*. For any pair of rows (i_1, i_2), the multiset

$$\{M[i_1, j] * M[i_2, j]^{-1}\colon\ j = 1, 2, \ldots, s\lambda\}$$

contains the identity element of G exactly λ times. If M has the additional property that, for any $\alpha, \beta \in G$ and any two rows i_1, i_2,

$$|\ \{j\colon\ M[i_1, j] = \alpha \text{ and } M[i_2, j] = \beta\}\ | = |\ \{j\colon\ M[i_1, j] = \beta \text{ and } M[i_2, j] = \alpha\}\ |,$$

we say that the PBD has an *obverse matrix* M over G.

THEOREM 2.2 *Let s be a prime power. If there is an ℓ by w PBD(v, λ) with $w \leq s \leq \lfloor \frac{\ell}{\lambda} \rfloor$, then there exists a $(s, v; \ell - \lambda)$-difference matrix over the additive group of GF(s).*

Proof. Let $(X, \mathcal{B})$ be an ℓ by w PBD(v, λ) and let M, over the additive group of $GF(s)$, be any matrix belonging to $(X, \mathcal{B})$. Choose a primitive root α of GF(s) and set $D' = [M, \alpha M, \alpha^2 M, \ldots, \alpha^{s-2} M]$. If in rows i_1, i_2 of column j of M the difference is nonzero, then every nonzero difference occurs in rows i_1, i_2 in column j of exactly one $\alpha^h M$. Hence the vector difference of any pair of rows of D' contains each nonzero element of GF(s) exactly $\ell - \lambda$ times and 0 exactly $\lambda(s - 1)$ times. The conditions guarantee that $\ell - \lambda \geq \lambda(s - 1)$. Thus $m = \ell - \lambda s = \ell - \lambda - \lambda(s - 1) \geq 0$ and we set Z to be the v by m zero matrix. Therefore

$$D = [M, \alpha M, \alpha^2 M, \ldots, \alpha^{s-2} M, Z]$$

is a $(s, v; \ell - \lambda)$-difference matrix over the additive group of GF(s). ∎

For example, the 7 by 3 PBD(7, 2) given above yields a (3, 7; 5)-difference matrix over $\mathbb{Z}_3$ and the 7 by 5 PBD(7, 1) yields a (5, 7; 6)-difference matrix.

LEMMA 2.3 *Let $v = 1 + nk$ be a prime power. Then*

(i) *there is v by $n + 1$ PBD$(v, k - 1)$ $(X, \mathcal{B})$;*

(ii) *if $v = 2^m$ for some m the PBD in* (i) *has an obverse matrix over any group G of order $s \geq n + 1$; and*

(iii) *if v is odd and k is even the PBD$(v, k-1)$ in* (i) *has an obverse matrix over any group G of order $s \geq n+1$.*

Proof. Let f be a primitive element of GF(v) and set $g = f^n$. $K = \{1, g, g^2, \ldots, g^{k-1}\}$ is a subgroup of GF$(v)^*$ of order k. Let $\mathcal{P}$ be the parallel class consisting of $\{0\}$ and the n cosets of K in GF$(v)^*$ and define for all $x \in$ GF(v) the parallel class $\mathcal{P}_x$ by

$$\mathcal{P}_x = \mathcal{P} + x = \{B + x\colon\ B \in \mathcal{P}\}.$$

Then $\mathcal{B} = \cup_{x \in GF(v)} \mathcal{P}_x$ is preserved by the 2-transitive group $\{x \mapsto \alpha x + \beta\colon\ \alpha, \beta \in \text{GF}(v), \alpha \neq 0\}$. $\mathcal{B}$ is a v by $n+1$ PBD$(v, k-1)$.

Let G be a group of order $s \geq n+1$ and fix any injection $\Phi\colon \mathcal{P} \to G$ such that $\Phi(\{0\}) = e$, the identity of G. For each $x \in$ GF(v) and $B \in \mathcal{P}$ define $\Phi_x(B+x) = \Phi(B)$ and construct the v by $n+1$ matrix $M = M_k$ as in the definition. That is, the $[i, j]$–entry of M_k is $\Phi_j(B+j)$ where $j \in$ GF(v) and $i \in B \in \mathcal{P}$.

We claim that, subject to the conditions of (ii) and (iii), M_k is an obverse matrix. For (ii), the matrix M_1 as defined above is just the addition table of the field GF(2^m) and an easy induction on m shows that it is obverse. Consequently, because M_k is a refinement of M_1 (i.e. if $M_1[i_1, x] = M_1[i_2, x]$, then $M_k[i_1, x] = M_k[i_2, x]$), M_k is obverse.

For (iii), when $k = 2$ the parallel classes of the PBD form a one factorization of the complete graph K_v and thus the matrix M_2 is obverse. Again any refinement of M_2 to a matrix M_{2h} is also obverse and so for k even the PBD has an obverse matrix M_k. If k is odd, a simple parity argument shows that the PBD cannot possibly have an obverse matrix. ■

THEOREM 2.4 *If the ℓ by w PBD(v, λ) has an obverse matrix M over the additive group of GF(s) where s is an odd prime power such that $w \leq s \leq \lfloor \frac{\ell}{\lambda} \rfloor$, then there exists a $(s, v; \frac{\ell-\lambda}{2})$-difference matrix over $(GF(s), +)$.*

Proof. Choose a primitive root α of GF(s) and set

$$D' = [M, \alpha M, \alpha^2 M, \ldots, \alpha^{\frac{s-1}{2}-1} M]$$

Consider the pair of rows i_1, i_2 of M and let π be a permutation of order two on the ℓ columns of M such that if (a, b) occurs in column j on these rows, then (b, a) occurs in column $\pi(j)$ whenever $a \neq b$. If $a = b$, then $\pi(j) = j$. Now in column j of M, if the difference is a nonzero element $x \in$ GF(s), then every nonzero difference occurs in rows i_1, i_2 in column j or $\pi(j)$ of exactly one $\alpha^h M$. Hence the vector difference of any pair of rows of D' contains each nonzero element of GF(s) exactly $(\ell - \lambda)/2$ times and 0 exactly $\lambda(s-1)/2$ times. The conditions guarantee that $(\ell - \lambda)/2 \geq \lambda(s-1)/2$. Thus $m = (\ell - \lambda s)/2 = (\ell - \lambda)/2 - \lambda(s-1)/2 \geq 0$ and we set Z to be the v by m zero matrix. Therefore

$$D = [M, \alpha M, \alpha^2 M, \ldots, \alpha^{\frac{s-1}{2}-1} M, Z]$$

is a $(s, v; \frac{\ell-\lambda}{2})$-difference matrix over the additive group of GF(s). ■

COROLLARY 2.5 *Let $v = 1 + nk$ and s be prime powers, with $n + 1 \leq s \leq \lfloor \frac{v}{k-1} \rfloor$. Then there exist a $(s, v; v - k + 1)$-difference matrix over $(GF(s), +)$ and there also exists a $(s, v; \frac{v-k+1}{2})$-difference matrix over $(GF(s), +)$ if v or k is even, and s is odd.*

Proof. Apply Theorems 2.2 and 2.4 to Lemma 2.3. ∎

In particular Corollary 2.5 settles the existence of a $(3, 7; 5)$-difference matrix and of a $(4, 13; 10)$-difference matrix.

The PBD$(v, k - 1)$ used in Corollary 2 is a special kind of nearly resolvable design. In general a *nearly resolvable design* with parameters NRB(v, k) is a v by $1 + (v - 1)/k$ PBD$(v, k - 1)$ in which every nonsingleton block has exactly k points and $v = 1 + nk$ for some n. In [5] it is reported that:

- An NRB$(v, 3)$ exists if and only if $v \equiv 1 \bmod 3$.
- An NRB$(v, 4)$ exists if and only if $v \equiv 1 \bmod 4$.
- An NRB$(v, 5)$ exists for $v \equiv 1 \bmod 5$, except possibly for v in $\{46, 51, 141, 201\}$.
- An NRB$(v, 6)$ exists for $v \equiv 1 \bmod 6$, except possibly for v in $\{55, 145\}$.

Applying Theorem 2.2 with s equal to the width, we get:

COROLLARY 2.6 *For every prime power s and integer $3 \leq k \leq 6$, there is a $(s, 1 + (s - 1)k; 2 + (s - 2)k)$-difference matrix over $(GF(s), +)$ except possibly for $(s, k) = (11, 5)$, $(29, 5)$, $(41, 5)$, or $(25, 6)$.*

Actually, more can be obtained because we can employ any prime power s in the range $1 + \frac{v-1}{k} \leq s \leq \frac{v-1}{k} + \frac{v}{k(k-1)}$.

We can also use resolvable designs:

COROLLARY 2.6 *There is $(s, 3s; 3(s - 1)/2)$-difference matrix over the additive group of $GF(s)$ for all odd prime powers s.*

Proof. Let s be an odd prime power and set $n = (s - 1)/2$. Then there is a Kirkman triple system of order $3s = 3 + 6n$ (see [2]). This is a $1 + 3n$ by $1 + 2n$ PBD$(3s, 1)$. Hence by Theorem 2.2 the result follows. ∎

For example using Corollary 2.7 a $(5, 15; 6)$-difference matrix can be constructed; this improves the bound on the number of rows v of a $(5, v; 6)$-difference matrix from $10 \leq v \leq 30$ to $15 \leq v \leq 30$.

COROLLARY 2.8 *If there is a resolvable BIBD with parameters (b, v, r, k, λ), with $\frac{v}{k}$ a prime power, then there is a $(\frac{v}{k}, v; r - \lambda)$-difference matrix.*

Proof. A resolvable BIBD is an r by v/k PBD(v, λ). Now

$$\frac{r}{\lambda} = \frac{v-1}{k-1} \geq \frac{v}{k}$$

Table 1. Some resolvable (v, b, r, k, λ) designs that give new $(\frac{v}{k}, v; r-\lambda)$-difference matrices.

No. in [8]	(v, b, r, k, λ)	$(\frac{v}{k}, v; r-\lambda)$-difference matrix
14	(15,35,7,3,1)	(5, 15; 6)-difference matrix
151	(65,208,16,5,1)	(13, 65; 15)-difference matrix
219	(39,247,19,3,1)	(13, 39; 18)-difference matrix
279	(85,357,21,5,1)	(17, 85; 20)-difference matrix

since in a BIBD $v > k$ and $\lambda(v-1) = r(k-1)$. Therefore by Theorem 2.2 a $(\frac{v}{k}, v; r-\lambda)$-difference matrix exists. ■

Several resolvable BIBD exist with which new difference matrices can be constructed using Corollary 2.8. Some examples are given in Table 1.

3. Two Constructions

In this section we describe two constructions which produce difference matrices over arbitrary groups.

THEOREM 3.1 *If an $OA_\lambda(k, n)$ exists with at least λ constant columns, then, over any group G of order $n+1$, a $(n+1, k; \lambda(n-1))$-difference matrix exists.*

Proof. Let G be any group of order $n+1$ with identity element e. Let A be an $\mathrm{OA}_\lambda(k, n)$, with entries from $X = G \setminus \{e\}$. In this array every ordered pair of symbols of X occurs λ times in each ordered pair of rows. Hence in the multiset

$$\{A[i_1, j] * A[i_2, j]^{-1}\colon\ j = 1, 2, \ldots, s\lambda\}$$

for any pair of rows (i_1, i_2) e occurs λn times and each nonidentity element of G occurs $\lambda(n-1)$ times. Deleting λ of the constant columns constructs the desired difference matrix. ■

A (10, 27; 24)-difference matrix exists by Theorem 3.1, using an $\mathrm{OA}_3(27, 9)$.

THEOREM 3.2 *If v is a prime power and $v = 1 + nk$ for nonnegative integers n and k with $n \geq k-2 \geq 0$, then for any group G of order $n+1$ there is a $(n+1, v; 2+(n-1)k)$-difference matrix over G.*

Proof. Let f be a primitive element of GF(v) and set $g = f^n$. Then the order of g in GF$(v)^*$ is k. Hence, $B = \{g^j\colon\ 0 \leq j < k\}$ is a subgroup of GF$(v)^*$ of order k and $B_i = f^{i-1}B, i = 1, 2, 3, \ldots, n$ is a complete set of cosets. Let $B_0 = \{0\}$. Define the v by v matrix M indexed by GF(v) and with entries in $G = \{a_0 = e, a_1, a_2, \ldots a_n\}$ by:

$$M[\alpha, \beta] = a_i \text{ if and only if } \alpha + \beta \in B_i.$$

Now set π to be the permutation of the elements of G given by $\pi = (e)(a_1, a_2, a_3, \ldots, a_n)$ and define $\pi(M)$ to be the v by v matrix with $[i, j]$-th entry $\pi(M[i, j])$. Finally set Z to be the v by $(n-k+2)$ constant matrix containing e and define the v by $(nv+n-k+2)$ matrix D to be

$$D = [M, \pi(M), \pi^2(M), \ldots, \pi^{n-1}(M), Z]$$

We claim that D is difference matrix over G. Let $\alpha, \beta \in \mathrm{GF}(v), \alpha = f^\ell \ (1 \le \ell \le n-1)$, and consider any $i, j \in \mathrm{GF}(v)$. If $a_u = M[\alpha \cdot i + \beta, \alpha \cdot j - \beta]$, then $f^\ell(i+j) = \alpha(i+j) = \alpha \cdot i + \beta + \alpha \cdot j - \beta \in B_u = f^{u-1}B$. Hence $i + j \in f^{u-\ell-1}B = B_{u-\ell}$, which implies $M[i, j] = \pi^{-\ell}(a_u)$. Now fix a pair of rows i_0, i_0' of M and consider any other two rows i_1, i_1'. Since the group $H = \{x \mapsto \alpha x + \beta \colon \alpha, \beta \in \mathrm{GF}(v)\}$ acts sharply 2–transitively on $\mathrm{GF}(v)$, there is a unique $\alpha, \beta \in \mathrm{GF}(v)$ such that $\alpha \cdot i_0 + \beta = i_1$ and $\alpha \cdot i_0' + \beta = i_1'$. The above argument shows that the pair $(M[i_1, \alpha \cdot j - \beta], M[i_1', \alpha \cdot j - \beta]) = \pi^\ell(M[i_0, j], M[i_0', j])$. Hence $(M[i_1, \alpha \cdot j - \beta], M[i_1', \alpha \cdot j - \beta])$ and $(M[i_0, j], M[i_0', j])$ belong to the same orbit of π on ordered pairs. In particular if the orbit of (a_i, a_j) under π is represented $N_{i,j}$ times in some ordered pair of rows of M it is represented exactly $N_{i,j}$ times in every pair of rows. The value of $N_{i,j}$ can now be calculated. Let (x, y) be in the orbit of (a_i, a_j) under π.

Case 1. $i \neq 0, j \neq 0$.

In this case there are in column α of M exactly k^2 ordered pairs of rows that have (x, y). The length of the orbit of (a_i, a_j) under π is n and so in column α there are exactly nk^2 ordered pairs of rows that have an orbit representative of the orbit of (a_i, a_j) under π. Summing over all columns and dividing by the number of ordered pairs of rows, we find $N_{i,j} = k$.

Case 2. $i = 0, j \neq 0$ or $i \neq 0, j = 0$.

In this case there are in column α of M exactly k ordered pairs of rows that have (x, y). Following exactly the same argument as in case 1 we have in this case $N_{i,j} = 1$.

This regularity on M shows that in any pair of rows in D every ordered pair (a_i, a_j) occurs k times if neither i nor j is 0 and occurs once otherwise. Hence in any pair of rows each nonidentity difference occurs $k(n-1)+2$ times. A pair (a_i, a_j) has difference e if and only if $i = j$. If the pair is in column α and rows (β_1, β_2) of M this means that $\beta_1, \beta_2 \in B_i - \alpha$. But $\{B_i + \gamma \colon \gamma \in \mathrm{GF}(v)\}$ is a set system invariant under the 2-transitive group H and is hence a 2–$(v, k, k-1)$ design. So the identity of G occurs as a difference in any particular pair of rows of D exactly $n(k-1) + n - k + 2 = k(n-1) + 2$ times. ■

Some examples of this construction appear in Table 2.

4. Concluding Remarks

We have developed a number of constructions here for difference matrices, but the number of rows obtained is relatively small. Nevertheless, the results obtained often improve upon

Table 2. Some new difference matrices over arbitrary groups.

A (6, 11; 10)-difference matrix over the group $\mathcal{A} = \{a_0 = e, a_1, \ldots, a_5\}$ is constructed by cyclically shifting the following five columns (shown transposed) into 55 columns and appending five columns of all zeros.

$$\begin{matrix} a_0 & a_1 & a_2 & a_4 & a_3 & a_5 & a_5 & a_3 & a_4 & a_2 & a_1 \\ a_0 & a_2 & a_3 & a_5 & a_4 & a_1 & a_1 & a_4 & a_5 & a_3 & a_2 \\ a_0 & a_3 & a_4 & a_1 & a_5 & a_2 & a_2 & a_5 & a_1 & a_4 & a_3 \\ a_0 & a_4 & a_5 & a_2 & a_1 & a_3 & a_3 & a_1 & a_2 & a_5 & a_4 \\ a_0 & a_5 & a_1 & a_3 & a_2 & a_4 & a_4 & a_2 & a_3 & a_1 & a_5 \end{matrix}$$

A (4, 13; 10)-difference matrix over the group $\mathcal{A} = \{a_0 = e, a_1, a_2, a_3\}$ is constructed by cyclically shifting the following three columns (shown transposed) into 39 columns and appending one column of all zeros.

$$\begin{matrix} a_0 & a_1 & a_2 & a_2 & a_3 & a_1 & a_3 & a_3 & a_1 & a_3 & a_2 & a_2 & a_1 \\ a_0 & a_2 & a_3 & a_3 & a_1 & a_2 & a_1 & a_1 & a_2 & a_1 & a_3 & a_3 & a_2 \\ a_0 & a_3 & a_1 & a_1 & a_2 & a_3 & a_2 & a_2 & a_3 & a_2 & a_1 & a_1 & a_3 \end{matrix}$$

the best available results in the literature. To see what effect the results developed have upon the bounds on the number of rows, we tabulate in Tables 3 and 4 the best lower bounds on the number v of rows in a $(s, v; \lambda)$-difference matrix for $1 \leq s \leq 32$ and $1 \leq \lambda \leq 30$. A key is provided to interpret the authority for each entry, which is given as a single letter superscript on the entry. When the superscript is omitted, the entry is obtained by the addition or multiplication construction stated in the introduction.

Key For Tables 3 and 4	
blank	obtained by addition or multiplication
d	de Launey [4]
g	generalized Hadamard matrix; see [3]
h	Hadamard matrix; see [2]
j	Jungnickel [7]
m	Theorem 3.2
n	Theorem 3.1
p	Theorem 2.2
q	Theorem 2.4
s	sporadic example; see [1] and [2]
t	tensor product; see [9]

We do not include the upper bounds, as for the most part they can be calculated easily. When $s \equiv 2 \pmod 4$ and $\lambda \equiv 1 \pmod 2$, a $(s, v; \lambda)$-difference matrix exists only for $v \leq 2$ (see, for example, [2]). In the remaining cases, $v \leq s\lambda$ always provides an upper bound [7]; when equality holds, the difference matrix is a *generalized Hadamard matrix*.

Table 3. Lower Bounds on Numbers of Rows in $(s, k; \lambda)$-difference matrix.

$s \setminus \lambda$	1	2	3	4	5	6	7	8	9	10	11	12	13	14	15
2	2^g	4^h	2	8^h	2	12^h	2	16^h	2	20^h	2	24^h	2	28^h	2
3	3^g	6^g	9^g	12^g	7^m	18^g	9	24^g	27^g	30^g	9	36^g	12	18	21^d
4	4^g	8^g	12^g	16^g	8	12	12	32^g	36^g	13^m	12	28^j	16	56^g	12
5	5^g	10^g	7^q	20^g	25^g	15^p	17^q	20	20	50^g	17	25^d	20	20	33^j
6	2	6^s	2	6	2	6	2	10^j	2	11^m	2	8	2	16^m	2
7	7^g	14^g	7	28^g	11^q	15^p	49^g	28	28^d	15	28	28	28	98^g	28
8	8^g	16^g	8	32^g	8	16	56^g	64^g	21^p	16	32	32	21	56	56
9	9^g	18^g	27^g	36^g	18	54^g	27	72^g	81^g	36	27	108^g	36	54	54
10	2	4	2	8^s	2	5	2	10^s	2	5	2	10	2	5	2
11	11^g	22^g	11	44^g	11	22	11	44	19^q	22	121^g	44	22	22	44
12	6^s	6	6	6	6	8	6	8	9	11^j	6	15^j	6	8	8
13	13^g	26^g	13	52^g	13	26	13	52	19^q	26	23^q	52	169^g	27^p	65^p
14	2	5^s	2	7	2	7	2	8	2	7	2	13^j	2	7	2
15	5^s	7^s	7^j	7	7	9	7	10	7	7	7	10	7	9	13^j
16	16^g	32^g	16	64^g	16	32	16	128^g	16	32	16	64	16	32	65^p
17	17^g	34^g	17	68^g	17	34	17	68	19^q	34	23^q	68	27^q	34	31^q
18	2	6^s	2	8	2	9	2	9	2	9	2	12	2	9	2
19	19^g	38^g	19	76^g	19	38	19	76	19	38	23^q	76	27^q	38	31^q
20	4	8^s	5	10^s	5	10	5	13^j	7	10	7	12	7	10	12
21	6^s	6	7	7	7	9	7	12	7	7	7	14	7	9	9
22	2	4	2	8	2	11	2	11	2	11	2	12	2	11	2
23	23^g	46^g	23	92^g	23	46	23	92	23	46	23	92	27^q	46	31^q
24	6^s	6	8	8	7	9	8	12	8	8	8	16	8	9	8
25	25^g	50^g	25	100^g	125^g	50	50	100	100	125	50	100	100	100	125
26	2	4	2	8	2	12	2	13	2	13	2	13	2	13	2
27	27^g	54^g	81^g	108^g	54	162^g	81	108	243^g	108	81	324^g	108	108	162
28	5^s	7	7	8	7	12	7	14	7	8	7	12	7	12	11
29	29^g	58^g	29	116^g	29	58	29	116	29	58	29	116	29	58	31^q
30	2	5	2	6	2	6	2	7	2	6	2	8	2	6	2
31	31^g	62^g	31	124^g	31	62	31	124	31	62	31	124	31	62	31
32	32^g	64^g	32	128^g	32	64	32	256^g	32	64	32	128	32	64	32

Nonexistence of generalized Hadamard matrices for certain choices of s and λ (see, for example, [3]) reduces the upper bound to $v \leq s\lambda - 1$.

Acknowledgements

Charles J. Colbourn's research was supported by NSERC Canada grant A0579. Donald L. Kreher's research was supported by National Security Agency grant MDA904–92–H–3036. This work was completed while the authors held Raybould Fellowships at the University of Queensland, Australia. Special thanks to Anne Street, Liz Billington and the other members of the Department of Mathematics there. The authors were also supported in part by NSF Grant DMS-9402637. We especially thank Neil Sloane for suggesting the problem and for his encouragement, and Dieter Jungnickel for some very helpful comments.

Table 4. Lower Bounds on Numbers of Rows in $(s, k; \lambda)$-difference matrix (cont'd).

$s \setminus \lambda$	16	17	18	19	20	21	22	23	24	25	26	27	28	29	30
2	32^h	2	36^h	2	40^h	2	44^h	2	48^h	2	52^h	2	56^h	2	60^h
3	48^g	24	54^g	27	30	27	30	21	72^g	27	30	81^g	42^d	27	90^t
4	64^g	32	36	13	28	28	32	36	36^t	36	32	108^g	56	28	56
5	80^g	35^d	90^g	20	100^g	25	25	25	60^d	125^g	50	35	50	25	75^t
6	20^j	2	10	2	25^j	2	10	2	15^j	2	31^m	2	16	2	16
7	37^q	28	63^d	28	28	49	28	37	28	49	182^g	31^m	196^g	28	91^d
8	128^g	21	32	32	32	56	56	56	64	32	32	32	56	56	99^d
9	72	72	162^g	36	72	81	54	54	108	72	72	243^g	72	72	108
10	18^j	2	19^m	2	10	2	8	2	27^n	2	10	2	10	2	12
11	44	37^q	22	44	44	37	242^g	44	121^d	37	44	44	61^q	37	44
12	12	8	12	9	22^j	9	23^m	8	16	9	11	12	12	9	12
13	52	52	39^p	52	52	52	39	52	52	52	338^g	52	169^d	52	65
14	14	2	7	2	14	2	7	2	26^j	2	27^m	2	14	2	11
15	12	7	10	7	12	9	9	10	18	7	10	10	29^m	9	18
16	256^g	33^p	39^p	64	64	45^p	39	65	128	45	39	64	64	45	65
17	272^g	289^g	39^p	34	85^p	68	39	34	68	68	39	34	68	68	39
18	17^j	2	12	2	18	2	9	2	18	2	9	2	18	2	20
19	76	31	39^p	361^g	85^p	45^p	39	76	76	45	39	76	76	61^q	39
20	16	8	19^j	10	19^j	12	10	12	15	10	13	12	16	12	12
21	14	7	14	7	14	15^j	9	9	18	7	12	9	14	12	14
22	16	2	11	2	20	2	11	2	22	2	11	2	22	2	11
23	92	31	46	31	92	43^q	46	529^g	92	46	46	92	92	46	63^p
24	16	8	16	8	16	9	23^j	8	18	8	12	9	16	9	16
25	100	100	100	100	125	100	100	100	100	625^g	100	100	100	100	125
26	16	2	13	2	20	2	13	2	25^j	2	13	2	26	2	13
27	108	108	243	108	108	243	108	108	324	108	108	729^g	108	108	243
28	16	7	14	8	13	12	12	11	15	8	27^j	12	19^j	12	12
29	116	31	58	31	116	43^q	58	47^q	116	47	58	47	116	841^g	63^p
30	10	2	7	2	10	2	6	2	9	2	7	2	29^j	2	7
31	124	31	62	31	124	43^q	62	47^q	124	47	62	47	124	59^q	63^p
32	512^g	32	64	32	128	32	64	32	256	32	64	32	128	32	64

References

1. R. J. R. Abel and Y. W. Cheng, Some new MOLS of order $2^n p$ for p a prime power, *Austral. J. Combin.*, 10 (1994) pp. 175–186.
2. T. Beth, D. Jungnickel, and H. Lenz, *Design Theory*, Cambridge University Press (1986).
3. W. de Launey, A survey of generalized Hadamard matrices and difference matrices $D(k, \lambda; G)$ with large k, *Util. Math.*, 30 (1986) pp. 5–29.
4. W. de Launey, On difference matrices, transversal designs, resolvable transversal designs, and large sets of mutually orthogonal F-squares, *J. Stat. Plan. Infer.*, 16 (1987) pp. 107–125.
5. S. Furino, Existence results for near resolvable designs, *J. Comb. Designs*, 3 (1995) pp. 101–113.
6. A. S. Hedayat, N. J. A. Sloane, and J. Stufken, *Orthogonal Arrays*, to appear.
7. D. Jungnickel, On difference matrices, resolvable transversal designs and generalized Hadamard matrices, *Math. Z.*, 167 (1979) pp. 49–60.
8. R. Mathon and A. Rosa, Tables of parameters of BIBDs with $r \leq 41$ including Existence, Enumeration, and Resolvability Results, *Ann. Disc. Math.*, 26 (1985) pp. 275–308.
9. S. S. Shrikhande, Generalized Hadamard matrices and orthogonal arrays of strength 2, *Canad. J. Math.*, 16 (1964) pp. 131–141.

Designs, Codes and Cryptography, 9, 71–77 (1996)
© 1996 Kluwer Academic Publishers, Boston. Manufactured in The Netherlands.

Hadamard Matrices and Dihedral Groups

HIROSHI KIMURA
Department of Mathematics, Ehime University, Matsuyama 790–77, Japan

Communicated by: V. Tonchev

Received November 3, 1994; Accepted August 16, 1995

Abstract. Let D_{2p} be a dihedral group of order $2p$, where p is an odd integer. Let ZD_{2p} be the group ring of D_{2p} over the ring Z of integers. We identify elements of ZD_{2p} and their matrices of the regular representation of ZD_{2p}. Recently we characterized the Hadamard matrices of order 28 ([6] and [7]). There are exactly 487 Hadamard matrices of order 28, up to equivalence. In these matrices there exist matrices with some interesting properties. That is, these are constructed by elements of ZD_6. We discuss relation of ZD_{2p} and Hadamard matrices of order $n = 8p + 4$, and give some examples of Hadamard matrices constructed by dihedral groups.

Keywords: Dihedral groups, Hadamard matrices

1. Observation of Hadamard Matrices of Order 28

A Hadamard matrix H of order n is an $n \times n$ matrix of $\pm 1's$ with $HH^t = nI$. It is well known that n is necessarily 1, 2 or a multiple of four. We say that two matrices M_1 and M_2 of same size are equivalent if there exists a signed permutation g of rows and columns of M_1 with $M_1^g = M_2$. A matrix which is equivalent to a Hadamard matrix is also a Hadamard matrix. We say that a set of four rows of H is a Hall set if the submatrix consisting of the four rows is equivalent to the following matrix:

$$\begin{pmatrix} + & + & + & + & J_m & J_m & J_m & J_m \\ + & + & - & - & J_m & J_m & -J_m & -J_m \\ + & - & + & - & J_m & -J_m & J_m & -J_m \\ + & - & - & + & -J_m & J_m & J_m & -J_m \end{pmatrix}, \tag{1}$$

where J_m is the all $1's$ row vector of dimension $(n-4)/4$.

Let J be the square matrix of all $1's$. In this paper we say that $(H + J)/2$ is a Hadamard matrix and identify this with H. That is, a Hadamard matrix H of order n is an $n \times n$ matrix of 0, $1's$ such that the number of 1 contained in the sum of every two rows of H equals $n/2$.

The equivalence classes of Hadamard matrices of order ≤ 28 have been determined by Hall, Ito-Leon-Longyear and the author ([2], [3], [4], [5], [6] and [7]). There are exactly 486 inequivalent Hadamard matrices of order 28 with Hall sets.

Let $D_{2p} = \langle x, y \colon |x| = p, |y| = 2, yxy = x^{-1} \rangle$ be a dihedral group of order $2p$, where p is an odd integer. Let ZD_{2p} be the group ring of D_{2p} over the ring Z of integers. It is clear that ZD_{2p} is isomorphic to the factor ring of the non-commutative ring of two variable X, and Y over Z by the ideal generated by $X^p - 1$, $Y^2 - 1$ and $YXY - X^{p-1}$. We identify the natural images of X and Y in the factor ring with x and y, respectively. Furthermore we identify elements of ZD_{2p} with the corresponding matrices in the regular representation of ZD_{2p}. If $a = \sum_{g \in G} a_g g$ is an element of the group ring and A is its matrix representation, then A^T is the matrix that represents $\sum_{g \in G} a_g g^{-1}$.

The Hadamard matrix H_{450} of order 28 in [8] is equivalent to the following matrix

$$\begin{pmatrix}
1111 & 111 & 111 & 111 & 111 & 111 & 111 & 111 & 111 \\
1100 & 111 & 111 & 111 & 111 & 000 & 000 & 000 & 000 \\
1010 & 111 & 111 & 000 & 000 & 111 & 111 & 000 & 000 \\
1001 & 000 & 000 & 111 & 111 & 111 & 111 & 000 & 000 \\
\\
1110 & 110 & 000 & 101 & 001 & 101 & 010 & 101 & 100 \\
1110 & 011 & 000 & 110 & 100 & 110 & 001 & 110 & 010 \\
1110 & 101 & 000 & 011 & 010 & 011 & 100 & 011 & 001 \\
1110 & 000 & 101 & 010 & 110 & 001 & 110 & 100 & 110 \\
1110 & 000 & 110 & 001 & 011 & 100 & 011 & 010 & 011 \\
1110 & 000 & 011 & 100 & 101 & 010 & 101 & 001 & 101 \\
\\
1101 & 010 & 110 & 110 & 000 & 101 & 100 & 010 & 101 \\
1101 & 001 & 011 & 011 & 000 & 110 & 010 & 001 & 110 \\
1101 & 100 & 101 & 101 & 000 & 011 & 001 & 100 & 011 \\
1101 & 101 & 001 & 000 & 101 & 100 & 110 & 110 & 001 \\
1101 & 110 & 100 & 000 & 110 & 010 & 011 & 011 & 100 \\
1101 & 011 & 010 & 000 & 011 & 001 & 101 & 101 & 010 \\
\\
1011 & 010 & 101 & 010 & 011 & 110 & 000 & 101 & 001 \\
1011 & 001 & 110 & 001 & 101 & 011 & 000 & 110 & 100 \\
1011 & 100 & 011 & 100 & 110 & 101 & 000 & 011 & 010 \\
1011 & 110 & 001 & 011 & 001 & 000 & 101 & 010 & 110 \\
1011 & 011 & 100 & 101 & 100 & 000 & 110 & 001 & 011 \\
1011 & 101 & 010 & 110 & 010 & 000 & 011 & 100 & 101 \\
\\
1000 & 101 & 100 & 010 & 101 & 101 & 001 & 001 & 111 \\
1000 & 110 & 010 & 001 & 110 & 110 & 100 & 100 & 111 \\
1000 & 011 & 001 & 100 & 011 & 011 & 010 & 010 & 111 \\
1000 & 100 & 110 & 110 & 001 & 010 & 110 & 111 & 010 \\
1000 & 010 & 011 & 011 & 100 & 001 & 011 & 111 & 001 \\
1000 & 001 & 101 & 101 & 010 & 100 & 101 & 111 & 100
\end{pmatrix} \tag{2}$$

This Hadamard matrix contains the following submatrix of size 24:

$$\begin{pmatrix}
A & B & C & D \\
\overline{B} & A & D & \overline{C} \\
\overline{C} & \overline{D} & A & B \\
D & \overline{C} & B & \overline{A}
\end{pmatrix}, \tag{3}$$

where A, B, C and D are submatrices of size 6, $\overline{A} = J - A$, $\overline{B} = J - B$, $\overline{C} = J - C$ and $\overline{D} = J - D$. Furthermore A, B, C and D can be considered as elements of ZD_{2p}, with $p = 3$.

Thus the following problem arises.

PROBLEM 1 *Find elements A, B, C and D of ZD_{2p} such that*

$$\begin{pmatrix} 1 & 1 & 1 & 1 & J_{2p} & J_{2p} & J_{2p} & J_{2p} \\ 1 & 1 & 0 & 0 & J_{2p} & J_{2p} & 0 & 0 \\ 1 & 0 & 1 & 0 & J_{2p} & 0 & J_{2p} & 0 \\ 1 & 0 & 0 & 1 & 0 & J_{2p} & J_{2p} & 0 \\ J_{2p}^t & J_{2p}^t & J_{2p}^t & 0 & A & B & C & D \\ J_{2p}^t & J_{2p}^t & 0 & J_{2p}^t & \overline{B} & A & D & \overline{C} \\ J_{2p}^t & 0 & J_{2p}^t & J_{2p}^t & \overline{C} & \overline{D} & A & B \\ J_{2p}^t & 0 & 0 & 0 & D & \overline{C} & B & \overline{A} \end{pmatrix} \tag{4}$$

is a Hadamard matrix.

In (4) A is a sum of $p-1$ elements, B, C, and D are sums of p elements of D_{2p}, respectively, since the fifth row is orthogonal to the first four rows. It is trivial that the matrix in Problem 1 is a Hadamard matrix if and only if the following conditions are satisfied:

CONDITION 2

$$AA^t + BB^t + CC^t + DD^t = (2p+1)I + (2p-2)J, \tag{5}$$

$$A\overline{B}^t + BA^t + CD^t + D\overline{C}^t = (2p-1)J, \tag{6}$$

$$A\overline{C}^t + B\overline{D}^t + CA^t + DB^t = (2p-1)J \tag{7}$$

and

$$AD^t + B\overline{C}^t + CB^t + D\overline{A}^t = 2pJ, \tag{8}$$

where $J = \sum_{g \in D_{2p}} g$ and $\overline{A} = J - A$.

The following proposition is very useful for a computer search.

PROPOSITION 3 *If (A, B, C, D) satisfies Condition 2, then (A, C, B, D), $(A, B, \overline{D}, \overline{C})$ and $(A^t, \overline{B}^t, \overline{C}^t, D^t)$ also satisfy Condition 2.*

Proof. The first two cases come from permutations of first four rows and columns of (4). The last is trivial since the transpose of a Hadamard matrix is a Hadamard matrix. ■

For $p = 3$ the following elements:

$$\begin{cases} A = & 1+x, \\ B = & 1+x^2+x^2y, \\ C = & 1+x^2+xy, \\ D = & 1+x^2+y, \end{cases} \tag{9}$$

and

$$\begin{cases} A = & 1+x, \\ B = & 1+y+x^2y, \\ C = & 1+xy+x^2y, \\ D = & 1+y+xy, \end{cases} \tag{10}$$

satisfy Condition 2.

2. Conditions for A, B, C, and D

Let ϕ: $ZD_{2p} \to Z(D_{2p}/\langle x \rangle)$ be the canonical homomorphism given by $g \mapsto \overline{g}$ and let $\phi(A) = a_1\overline{1} + a_2\overline{y}$, $\phi(B) = b_1\overline{1} + b_2\overline{y}$, $\phi(C) = c_1\overline{1} + c_2\overline{y}$ and $\phi(D) = d_1\overline{1} + d_2\overline{y}$. Then since $y^{-1} = y$ the image of equation (5) under ϕ is

$$\begin{aligned} &(a_1\overline{1} + a_2\overline{y})^2 + (b_1\overline{1} + b_2\overline{y})^2 + (c_1\overline{1} + c_2\overline{y})^2 + (d_1\overline{1} + d_2\overline{y})^2 \\ &\quad = (2p+1)\overline{1} + (2p-2)p\overline{1} + (2p-2)p\overline{y}. \end{aligned}$$

Simplifying this expression and equating coefficients of $\overline{1}$ and $\overline{y}$ yields the following proposition.

PROPOSITION 4 *The equations:*

$$\begin{cases} a_1 + a_2 = p - 1, \\ b_1 + b_2 = c_1 + c_2 = d_1 + d_2 = p, \\ a_1^2 + a_2^2 + b_1^2 + b_2^2 + c_1^2 + c_2^2 + d_1^2 + d_2^2 = 2p^2 + 1, \\ a_1a_2 + b_1b_2 + c_1c_2 + d_1d_2 = p(p-1) \end{cases} \tag{11}$$

must be satisfied.

For $p \leq 27$, we have the following parameters (see Figure 1).

We consider several cases. For example, suppose $p = 5$ and $n = 44$. In this case the following matrix is a submatrix of Hadamard matrix of order 44 which is the first row in (3):

$$(A, B, C, D) = \begin{pmatrix} 00110 & 01001 & 10000 & 01111 & 11001 & 01001 & 11001 & 01001 \\ 00011 & 10100 & 01000 & 10111 & 11100 & 10100 & 11100 & 10100 \\ 10001 & 01010 & 00100 & 11011 & 01110 & 01010 & 01110 & 01010 \\ 11000 & 00101 & 00010 & 11101 & 00111 & 00101 & 00111 & 00101 \\ 01100 & 10010 & 00001 & 11110 & 10011 & 10010 & 10011 & 10010 \\ 01001 & 00110 & 01111 & 10000 & 01001 & 11001 & 01001 & 11001 \\ 10100 & 00011 & 10111 & 01000 & 10100 & 11100 & 10100 & 11100 \\ 01010 & 10001 & 11011 & 00100 & 01010 & 01110 & 01010 & 01110 \\ 00101 & 11000 & 11101 & 00010 & 00101 & 00111 & 00101 & 00111 \\ 10010 & 01100 & 11110 & 00001 & 10010 & 10011 & 10010 & 10011 \end{pmatrix}. \tag{12}$$

Figure 1. a_i, b_i, c_i, d_i for $p \leq 27$.

p	a_1	a_2	b_1	b_2	c_1	c_2	d_1	d_2	p	a_1	a_2	b_1	b_2	c_1	c_2	d_1	d_2
3	2	0	1	2	1	2	1	2	19	12	6	9	10	9	10	9	10
5	2	2	1	4	3	2	3	2		10	8	7	12	11	8	9	10
7	4	2	5	2	3	4	3	4	21	12	8	13	8	11	10	11	10
9	6	2	5	4	5	4	5	4		12	8	9	12	9	12	9	12
	4	4	3	6	3	6	5	4		10	10	13	8	9	12	9	12
11	6	4	7	4	7	4	5	6	23	14	8	13	10	11	12	11	12
13	8	4	5	8	7	6	7	6		12	10	9	14	13	10	13	10
	6	6	9	4	7	6	7	6	25	14	10	15	10	11	14	13	12
	6	6	5	8	5	8	5	8		12	12	15	10	15	10	13	12
15	8	6	9	6	9	6	9	6		12	12	9	16	13	12	13	12
	8	6	5	10	7	8	7	8	27	16	10	15	12	15	12	13	14
17	10	6	7	10	7	10	9	8		14	12	17	10	13	14	13	14
	8	8	11	6	7	10	9	8		14	12	11	16	11	16	13	14

A, B, C and D express the following elements of ZD_{10}, respectively.

$$\begin{cases} A = & & +x^2 & +x^3 & & +xy & & & +x^4y, \\ B = & 1 & & & & +xy & +x^2y & +x^3y & +x^4y, \\ C = & 1 & +x & & +x^4 & +xy & & & +x^4y, \\ D = & 1 & +x & & +x^4 & +xy & & & +x^4y. \end{cases} \tag{13}$$

In this case the following assumption is also satisfied.

ASSUMPTION 5

$$A_1, A_2, B_1, B_2, C_1, C_2, D_1 \textit{ and } D_2 \textit{ are central elements in } ZD_{2p}. \tag{14}$$

In the general case we consider Problem 1 under this assumption. It seems that this assumption is not so strong.

THEOREM 6 *If* $A_1, A_2, B_1, B_2, C_1, C_2, D_1$ *and* D_2 *are central elements in* ZD_{2p}, *then Condition 2 simplifies to the following:*

$$A_1^2 + A_2^2 + B_1^2 + B_2^2 + C_1^2 + C_2^2 + D_1^2 + D_2^2 = (2p+1) + (2p-2)J_1, \tag{15}$$

$$A_1A_2 + B_1B_2 + C_1C_2 + D_1D_2 = (p-1)J_1, \tag{16}$$

where $J_1 = \sum_{g \in \langle x \rangle} g$.

Proof. Since $A_1^t = A_1$, $A_2^t = A_2$ and so on, it is easy to show that equation (5) is obtained from equations (15) and (16). Since $A\overline{B}^t = A(J-B)^t = A(J-B) = (a_1+a_2)J - AB$, we have

$$A\overline{B}^t + BA^t = (a_1+a_2)J = (p-1)J,$$

$$CD^t + D\overline{C}^t = (d_1+d_2)J = pJ.$$

This implies equation (6). Equations (7) and (8) are similarly obtained. ■

3. Examples of Hadamard Matrices

In this section we give some examples of Hadamard matrices for some p. Unfortunately we can not find a general solution of Problem 1. However the following examples show the success of this technique. For each example we write coefficients of elements in ZD_{2p}. For example, for $p=7$ (A_1: 0 0 1 1) represents $A_1 = x^2+x^3+x^{-3}+x^{-2}$ and (B_1: 1 1 0 1) represents $B_1 = 1+x^1+x^3+x^{-3}+x^{-1}$.

Example 1. $p=7$ and $n = 8\cdot 7+4 = 60$.

$$\begin{cases} A_1 : 0\ 011 & A_2 : 0\ 100 \\ B_1 : 1\ 101 & B_2 : 0\ 100 \\ C_1 : 1\ 010 & C_2 : 0\ 110 \\ D_1 : 1\ 100 & D_2 : 0\ 110 \end{cases}$$

Example 2. $p=9$ and $n = 8\cdot 9+4 = 76$.

$$\begin{cases} A_1 : 0\ 0111 & A_2 : 0\ 0100 \\ B_1 : 1\ 1001 & B_2 : 0\ 0101 \\ C_1 : 1\ 0101 & C_2 : 0\ 1100 \\ D_1 : 1\ 1010 & D_2 : 0\ 1100 \end{cases}$$

Example 3. $p=13$ and $n = 8\cdot 13+4 = 108$.

$$\begin{cases} A_1 : 0\ 10011\ 1 & A_2 : 0\ 01100\ 0 \\ B_1 : 1\ 00010\ 1 & B_2 : 0\ 11101\ 0 \\ C_1 : 1\ 01001\ 1 & C_2 : 0\ 01001\ 1 \\ D_1 : 1\ 10110\ 0 & D_2 : 0\ 10110\ 0 \end{cases}$$

Example 4. $p=17$ and $n = 8\cdot 17+4 = 140$.

$$\begin{cases} A_1 : 0\ 11010\ 001 & A_2 : 0\ 11010\ 001 \\ B_1 : 1\ 11101\ 001 & B_2 : 0\ 10000\ 110 \\ C_1 : 1\ 01000\ 110 & C_2 : 0\ 11111\ 000 \\ D_1 : 1\ 01101\ 001 & D_2 : 0\ 10111\ 000 \end{cases}$$

Example 5. $p = 41$ and $n = 8 \cdot 41 + 4 = 332$.

$$\begin{cases} A_1 & : 0 \ 10000 \ 10111 \ 01101 \ 10001 \\ A_2 & : 0 \ 01111 \ 01000 \ 10010 \ 01110 \\ B_1 & : 1 \ 01000 \ 11101 \ 10100 \ 11111 \\ B_2 & : 0 \ 10111 \ 00010 \ 01011 \ 00000 \\ C_1 & : 1 \ 10010 \ 10001 \ 00011 \ 11110 \\ C_2 & : 0 \ 10010 \ 10001 \ 00011 \ 11110 \\ D_1 & : 1 \ 10110 \ 01001 \ 11100 \ 10100 \\ D_2 & : 0 \ 10110 \ 01001 \ 11100 \ 10100 \end{cases}$$

Remark. We researched Example 5 under Assumption 5 and the assumptions $A_1 = J - 1 - A_2$, $B_1 = J - B_2$, $C_1 = 1 + C_2$ and $D_1 = 1 + D_2$. For $p = 5, 13$, and 37 there are Hadamard matrices satisfing the above assumptions. For $p = 53$, there is no Hadamard matrix satisfiying the assumptions.

References

1. M. Hall, Jr., *Combinatorial Theory*, Ginn (Blaisdell) Boston (1967).
2. M. Hall, Jr., Hadamard matrices of order 16, *J. P. L. Research Summary, 36–10*, Vol. 1 (1961) 21–26.
3. M. Hall, Jr., Hadamard matrices of order 20, *J. P. L. Technical Report*, (1965) 32–761.
4. N. Ito, J. S. Leon, and J. Q. Longyear, Classification of 3-(24,12,5) designs and 24-dimensional Hadamard matrices, *J. Combin. Theory(A)*, Vol. 27 (1979) pp. 289–306.
5. H. Kimura, New Hadamard matrix of order 24, *Graphs and Combin.*, Vol. 5 (1989) pp. 236–242.
6. H. Kimura, Classification of Hadamard matrices of order 28 with Hall sets, *Discrete Math.*, Vol. 128 (1994) pp. 257–268.
7. H. Kimura, Classification of Hadamard matrices of order 28, *Discrete Math.*, Vol. 133 (1994) pp. 171–180.
8. H. Kimura, A list of Hadamard matrices of order 28, unpublished manuscript.

Designs, Codes and Cryptography, 9, 79–83 (1996)
© 1996 Kluwer Academic Publishers, Boston. Manufactured in The Netherlands.

Self-Orthogonal Greedy Codes

LAURA MONROE monroe@uic.edu
Department of Mathematics and Computer Science, University of Illinois at Chicago, Chicago, IL 60657

Communicated by: V. Tonchev

Received November 4, 1994; Accepted December 1, 1995

Abstract. In this paper, we examine greedily generated self-orthogonal codes. We show that they are linear, and discuss several properties which they possess. We also look at specific examples of these codes.

Keywords: Greedy codes, self-orthogonal codes

1. Introduction

A binary greedy code of length n and designed distance d is generated by arranging the binary vectors of length n in some ordering. The first vector is selected for the code. Then, proceeding once through the ordering, a vector is selected if it differs from all previously chosen vectors in d or more places. Surprisingly, this set of vectors is linear, if the ordering is a B-ordering [1, 2, 3, 4, 5].

Greedy codes have been studied for several years, by Levenstein [4], Conway and Sloane [2], Pless and Brualdi [1], and others. Recently, A. J. van Zanten [7] generalized greedy codes by considering a set of difference vectors. A vector v is chosen for the code if all of the difference vectors $v \oplus c$ are in the set of permitted vectors, where c is in the code. He showed that if the set of permitted vectors satisfies a definite set of properties fixed upon before the greedy generation of the code, the code generated is linear. For example, the standard definition of greedy codes takes for its set of permitted vectors the set of all vectors having weight greater than or equal to a chosen minimum distance d. Van Zanten showed that the code is linear for any set of properties of the permitted set, not only this usual minimum weight property. Thus, his results generalize previous results.

One may also generate greedy codes with the additional stipulation that the codes generated be self-orthogonal. Linearity is not obvious in this case, since there is not an explicit set of permitted vectors existing prior to the greedy generation of the code.

Instead, the permitted set is generated as the greedy generation progresses through the B-ordering. This is true since each additional vector chosen imposes an additional restraint on further choices: the new choices must now be orthogonal to the last vector chosen, as well as all prior vectors. Thus, the permitted set changes as the algorithm progresses. It is not fixed at the beginning of the greedy generation, so the linearity of self-orthogonal greedy codes is not an immediate corollary of van Zanten's proof.

2. Linearity of Self-Orthogonal Greedy Codes

Definition. A *B-ordering* is an ordering of the length n vectors over the binary field obtained recursively from an ordered *B-basis* $\{b_1, \ldots, b_n)$, which can be any ordered basis of the length n binary vectors. The first vector in the B-ordering is the 0 vector, and the next is b_1. The B-ordering is then generated recursively, where if the first 2^{i-1} vectors of the ordering have been generated using B-basis elements $b_1, \ldots, b_{i-1}$, then the next 2^{i-1} vectors are generated by adding b_i to those vectors already produced, in order.

Definition. A *self-orthogonal greedy code* of length n and designed distance d is a greedy code generated with the additional restraint that the vectors must be orthogonal to themselves and each other, i.e., it is a code generated by first arranging the binary vectors of length n in an ordering, then selecting the first vector for the code, then, proceeding once through the ordering, selecting a vector if it both differs from all previously chosen vectors in d or more places and is orthogonal to itself and to all previously chosen vectors.

THEOREM 1 *Self-orthogonal greedy codes are linear, when generated from a B-ordering.*

Proof. Van Zanten showed in [7] that any greedy code generated from a B-basis is linear for any selection criteria P such that all of the difference vectors may be tested as to whether they satisfy P. The theorem then reduces to finding the appropriate P so that a self-orthogonal code with minimum distance d is produced. To assure the proper minimum distance, part of the selection criteria P must be that the distance vectors $x \oplus y$ have weights greater than or equal to d. The remaining part of P is obtained by considering the vectors in their B-ordering and generating a set S of vectors by requiring that the vectors in S differ from each other in at least d positions and that they are orthogonal to each other. P is then enlarged to include the stipulation that all difference vectors must be orthogonal to the vectors in S.

The self-orthogonal greedy code is thus generated via a two pass algorithm, the first pass to generate the criteria P, and the second to generate the actual self-orthogonal code. The only reason for using two passes is to prove the linearity of the self-orthogonal code generated. The code generated using the selection criteria P is the self-orthogonal greedy code and is linear, by van Zanten's proof. Note that the set S is actually the self-orthogonal greedy code itself. Thus, when actually generating a self-orthogonal greedy code, only one pass is necessary. ■

FACT *For any B-basis, the greedy self-orthogonal code for d odd is the same as the greedy self-orthogonal code for $d+1$.*

Proof. The code is linear, so all sums of vectors must be in the code. The code is self-orthogonal, so the weights of these sums must be even. This is the same as saying that all vectors in the code differ from each other in at least $d+1$ places. ■

Henceforth, when we refer to the chosen minimum distance d, it will be understood that d is even.

Definition. A *triangular B-basis* is one in which the ith basis element has 1 in the ith position, and 0 in all positions greater than i.

Standard greedy codes generated from triangular B-bases have been studied extensively, in [1] and in [5]. In general, they have been found to produce very good codes.

CONJECTURE *When the B-basis is triangular, for any d, greedy generation of a self-orthogonal code will eventually produce a self-dual code.*

We have generated such codes using the CRAY C-90, for small d. All data obtained at this point supports this conjecture. For $d = 4$, the (8, 4, 4) extended Hamming code is produced. For $d = 6$, the (22, 11, 6) baby Golay code is produced. For $d = 8$, the (24, 12, 8) extended Golay code is obtained.

Definition. A triangular B-basis is said to be *self-similar with period n* when the matrix whose rows are the basis elements in order has identical $n \times n$ matrices down the diagonal.

THEOREM 2 *When a self-dual $(n, n/2)$ code has been produced and the B-basis is triangular and self-similar with period n, the greedy self-orthogonal code of length mn will be a self-dual $(mn, mn/2)$ code which is the direct product of m copies of the original self-dual code.*

Proof is by induction on m. It is trivial for $m = 1$. Assume it is true for $m - 1$, and call the original self-dual code C. Then a self-dual $((m-1)n, (m-1)n/2)$ code $(m-1)C$ has been generated. Because $(m-1)C$ is self-dual, any additional vectors added to $(m-1)C$ would have to be identical to some element of $(m-1)C$ in the first $(m-1)n$ positions. The distance from the new vector to $(m-1)C$ would then be 0. Since the B-basis is self-similar, the next $2^{n/2}$ vectors would be C in positions from $(m-1)n + 1$ to mn.

THEOREM 3 *When B-basis is triangular and d is greater than 2 and not divisible by 4, the greedy self-orthogonal code cannot be a subcode of the greedy code generated with the same d and the same B-basis.*

Proof. Because the B-basis is triangular, the first element s_1 of the greedy self-orthogonal code is the vector consisting of d 1's. s_1 will also be the first vector of the greedy code. The second vector of either greedy code must have at least $d/2$ 1's to the left of the dth position, since otherwise it would either differ from s_1 or from the zero vector in less than d positions. The first such vectors occurring in any triangular B-ordering have exactly $d/2$ 1's to the left of the dth position. We construct a vector having exactly $d/2$ 1's to the left of the dth position which is eligible for inclusion in the greedy code and find the first such vector occurring in the B-ordering. This vector will then be the actual choice for the greedy code.

The vector must have at least $d/2$ 1's in the first d positions, since it must have weight at least d. It must also have at least $d/2$ 0's in the first d positions, since it must differ from s_1 in at least d positions. Hence it must have at most $d/2$ 1's in the first d positions. So the vector g_2 occurring first in the B-ordering which is a candidate for inclusion in the greedy code is the vector having first $d/2$ 1's, then some combination of $d/2$ 0's and $d/2$ 1's. Without loss of generality, we may permute the columns so that g_2 has the form $1\ldots1\,0\ldots0\,1\ldots1$. Since the greedy code is linear, the vector $g_2+s_1 = 1\ldots1\,1\ldots1\,0\ldots0$ is also in the code.

Now in the self-orthogonal case, the second vector in the code must have at least $d/2$ 1's to the left of the dth position, since otherwise it would differ either from s_1 or form the zero vector in less than d positions, and it also must have an even number of 1's in the first d positions, since the supports of the first and second vectors must intersect in an even number of positions. The second condition disqualifies any vector having exactly $d/2$ 1's to the left of the dth position since such a vector would be forced to have exactly $d/2$ 1's in the first d positions, as in the previous paragraph, which cannot happen, since $d/2$ is odd. So the second vector chosen must have at least $d/2+1$ 1's to the left of the dth position. The vector having first $d/2+1$ 1's, then $d/2+1$ 0's, then $d/2-1$ 1's meets both conditions, hence is a candidate for inclusion in the self-orthogonal greedy code. This vector occurs before any vector having more than $d/2+1$ 1's to the left of the dth position. So the second vector chosen must have exactly $d/2+1$ 1's to the left of the dth position. Let the chosen vector be called s_2. s_2 differs from g_2 and from g_2+s_1 by 1 to the left of the dth position. If s_2 were in the greedy code, then it would have to differ from g_2 by at least $d-1$ in the first d positions. But then it would differ from g_2+s_1 by at most 1 in the first d positions, and it would differ from g_2+s_1 by at most 2. (See illustration below.) So s_2 cannot be an element of the greedy code, since d is greater than 2.

$$
\begin{array}{ccccc}
 & d/2 & \quad\quad d & & \\
 & & 1\ldots\quad\quad\ldots 1 & & s_1 \\
 & 1\ldots1 & 0\ldots0 & 1\ldots1 & g_2 \\
 & 1\ldots1 & 1\ldots1 & 0\ldots0 & g_2+s_1 \\
1 & 1\ldots1 & {*}\ldots\quad\quad\ldots{*} & & s_2
\end{array}
$$

■

3. Some Examples

We have generated many examples of self-orthogonal binary greedy codes for small d using the CRAY C-90 supercomputer. The codes generated for the small d are quite interesting. As mentioned before, the $(8, 4, 4)$ extended Hamming code, the $(22, 11, 6)$ baby Golay code, and the $(24, 12, 8)$ extended Golay code are obtained for $d = 4, 6$, and 8, respectively, using a lexicographic basis. It is interesting that the self-orthogonal greedy code and the standard greedy code are exactly the same in the cases of $d = 4$ and $d = 8$, up to self-duality, at which point the two codes diverge (the self-orthogonal greedy code after this point becomes a direct product of the self-dual code, as in Theorem 2, but in the standard greedy code, non-doubly even vectors are produced).

Acknowledgements

The author was supported in part by grants from the U.S. Department of Education and the National Security Agency Grand MDA 904-91-H-0003. The author thanks the National Center for Supercomputing Applications, University of Illinois at Urbana-Champaign for the use of the CRAY C-90. The author would also like to thank V. Pless for many helpful discussions.

References

1. R. A. Brualdi and V. Pless, Greedy codes, *JCT(A)*, Vol. 64 (1993) pp. 10–30.
2. J. H. Conway and N. J. A. Sloane, Lexicographic codes: error correcting codes from game theory, *IEEE Trans. Inform. Theory*, Vol. IT-32 (1986) pp. 337–348.
3. D. Fon-Der-Flaass, A Note on greedy codes, to appear.
4. V.I. Levenstein, A Class of systematic codes, *Dokl. Akad. Nauk*, Vol. 1 (1960) pp. 368–371.
5. L. Monroe, Binary greedy codes, to appear in *Congressus Numerantium*, Vol. 100–104.
6. N. J. A. Sloane, Table of lower bounds on dmax (n, k) for linear codes over fields of order 3, The Handbook of Coding Theory (R. A. Brualdi, C. Huffman, and V. Pless, eds.), Elsevier Science Publishers, to appear.
7. N. J. A. Sloane, Table of lower bounds on dmax (n, k) for linear codes over fields of order 4, The Handbook of Coding Theory (R. A. Brualdi, C. Huffman, and V. Pless, eds.), Elsevier Science Publishers, to appear.
8. A. J. van Zanten, Lexicodes over fields of characteristic 2, to appear.

Designs, Codes and Cryptography, 9, 85–93 (1996)
© 1996 Kluwer Academic Publishers, Boston. Manufactured in The Netherlands.

On the Minimal Number of Even Submatrices of 0-1 Matrices

IOSIF PINELIS
Michigan Technological University, Department of Mathematical Sciences, Houghton, MI 49931-1295, USA

Communicated by: D. Jungnickel

Received November 1, 1994; Accepted October 10, 1995

Abstract. An asymptotic formula for the minimum possible number of even $p \times q$ submatrices of an $m \times n$ 0-1 matrix A is obtained. It is shown that if A is considered random and pq is even, then the distribution of the number of the even $p \times q$ submatrices of A is highly skewed to the right, the left endpoint of the distribution being very close to its mean, while its right endpoint is twice the mean. A relation to Turán numbers is indicated.

Keywords: 0-1 matrices, random matrices, even submatrices, Turán numbers

A matrix is called 0-1 if each entry of it is either 0 or 1. Let us call a 0-1 matrix even if the sum of all its entries is even, and odd otherwise. For a 0-1 matrix A, let $f(A; p, q)$ denote the number of all its even $p \times q$ submatrices. Let $F(m, n; p, q)$ stand for the minimum of $f(A; p, q)$ taken over all 0-1 $m \times n$ matrices A. Here and in what follows, m, n, p, and q are natural numbers.

The minimal number of even submatrices arises naturally in connection with the inequality

$$T(m+n, 5, 4) \leq \binom{m}{4} + \binom{m}{4} + F(m, n; 2, 2),$$

obtained by de Caen, Kreher, and Wiseman [1], where the *Turán number* $T(n, \ell, k)$ is the smallest possible number of k-subsets of an n-set such that every ℓ-subset contains a chosen k-set. See [1] also for the historical background.

THEOREM 1 *For every q, there exists a positive number $C(q)$, depending only on q and such that for all m, n, and p,*

$$F(m, n; p, 2q) \geq \frac{1}{2}\binom{m}{p}\binom{n}{2q}\left(1 - \frac{C(q)}{n^q}\right); \tag{1}$$

in particular,

$$F(m, n; p, 2) \geq \frac{1}{2}\binom{m}{p}\left\{\binom{n}{2} - \left\lfloor\frac{n}{2}\right\rfloor\right\}; \tag{2}$$

besides,

$$F(p,n;\,p,2)=\frac{1}{2}\left\{\binom{n}{2}-\left\lfloor\frac{n}{2}\right\rfloor\right\}. \tag{3}$$

Before we prove Theorem 1, let us make the following comments.

First, note that $F(m,n;\,p,q)=0$ if pq is odd. We therefore need to take care of $F(m,n;\,p,q)$ only in the case when at least one of the two numbers, p and q, is even.

Furthermore, let us consider an $m\times n$ 0-1 matrix A to be random, whose entries are statistically independent and take on values 0 and 1 with probability 1/2 each. Then (the minimal number) $F(m,n;\,p,q)$ cannot exceed the mathematical expectation of the distribution of $f(A;\,p,q)$, which is $\frac{1}{2}\binom{m}{p}\binom{n}{q}$, cf. [1, p. 279]:

$$F(m,n;\,p,q)\le\frac{1}{2}\binom{m}{p}\binom{n}{q}. \tag{4}$$

If, along with the minimal number $F(m,n;\,p,q)$ of even 0-1 submatrices, one introduces $G(m,n;\,p,q)$, the maximum of $f(A;\,p,q)$ taken over all 0-1 $m\times n$ matrices A, then $G(m,n;\,p,q)=\binom{m}{p}\binom{n}{q}$ for all m, n, p, and q (to see this, take A to be the zero matrix). When pq is odd, the distribution of $f(A;\,p,q)$ is symmetric with respect to its expectation, $\frac{1}{2}\binom{m}{p}\binom{n}{q}$; in other words, if $\nu(m,n;\,p,q;\,k)$ stands for the number of all 0-1 $m\times n$ matrices A satisfying $f(A;\,p,q)=k$, then $\nu(m,n;\,p,q;\,k)=\nu(m,n;\,p,q;\,N-k)$, where $N=\binom{m}{p}\binom{n}{q}$ and $k=0,\dots,N$; this is immediate because the transformation that converts each 0 entry of a 0-1 matrix A into 1, and vice versa, will take each even $p\times q$ submatrix of A to an odd one, and vice versa. Rather surprisingly, Theorems 1 and 2 show, in particular, that in the nontrivial case, when pq is even, the distribution of $f(A;\,p,q)$ is not only asymmetric, but highly skewed: the left endpoint $F(m,n;\,p,q)$ of the range of this distribution is very close to the expected value, $N/2$, while the right endpoint, $G(m,n;\,p,q)$, equals N.

For $p=2$, the above inequality (2) follows also from inequality (2) in [1], the latter deduced from some properties of Turán numbers.

Theorem 1 is nontrivial only if $m\ge p$ and $n\ge 2q$; these conditions will be assumed throughout the rest of the paper.

Theorem 1 is implied by the following two lemmata.

For any integer $k\in[0,n]$, set

$$g(n,k;\,q)=\sum_{j=0}^{q}\binom{k}{2j}\binom{n-k}{2q-2j};$$

note that, as usual, it is assumed that $\binom{n}{i}=0$ unless $0\le i\le n$.

LEMMA 1

$$F(p,n;p,2q) = \min\{g(n,k;q)\colon k = 0,\dots,n\}.$$

Proof. Consider any 0-1 $p \times n$ matrix A. Let $k = k(A)$ be the number of odd columns of A. Clearly, $f(A;p,2q) = g(n,k;q)$, which implies the lemma. ∎

LEMMA 2 *For every q, there exists a positive number $C(q)$, depending only on q and such that for all n and all $k = 0,\dots,n$,*

$$g(n,k;q) \geq \frac{1}{2}\binom{n}{2q}\left(1 - \frac{C(q)}{n^q}\right); \tag{5}$$

in particular,

$$g(n,k;1) \geq g(n,\lfloor n/2\rfloor;1) = \frac{1}{2}\left\{\binom{n}{2} - \left\lfloor\frac{n}{2}\right\rfloor\right\}. \tag{6}$$

Proof. One may assume that $k \leq n/2$. Using generating functions, it may be shown that

$$2g(n,k;q) = \binom{n}{2q} + \sum_{j=0}^{q}(-1)^j a_j, \tag{7}$$

where

$$a_j := \binom{n-2k}{2q-2j}\binom{k}{j}$$

[cf. line 7, page 134 of [2], where this identity is given with the obvious misprint: a minus in place of the plus before the second summation sign].

Observe that if $n - 2k \geq \sqrt{n} \geq 4q$, then

$$a_0 \geq \frac{1}{2^{2q}(2q)!}(n-2k)^{2q} \tag{8}$$

and

$$\sum_{j=1}^{q} a_j \leq \sum_{j=1}^{q}(n-2k)^{2q-2j}k^j \leq \sum_{j=1}^{q}(n-2k)^{2q-2j}n^j \leq q(n-2k)^{2q-2}n. \tag{9}$$

Now (7), (8), and (9) yield $g(n,k;q) \geq \frac{1}{2}\binom{n}{2q}$ if $n - 2k \geq C_1(q)\sqrt{n}$, where $C_1(q) = (q2^{2q}(2q)!)^{1/2}$, say.

Finally, if $0 \le n - 2k < C_1(q)\sqrt{n}$, then

$$\sum_{j=0}^{q} a_j \le \sum_{j=0}^{q} (n-2k)^{2q-2j} n^j \le C_2(q) n^q \le \frac{C(q)}{n^q}\binom{n}{2q},$$

where one could set $C_2(q) = (q+1)C_1(q)^{2q}$ and $C(q) = (2q)^{2q} C_2(q)$, because $\frac{1}{n^{2q}}\binom{n}{2q}$ increases in n if $n \ge 2q$.

This completes the proof of (5).

To prove (6), notice that

$$2g(n,k;1) = \binom{k}{2} + \binom{n-k}{2},$$

which is minimized by $k = \lfloor n/2 \rfloor$. ■

Theorem 1 now follows from the above lemmata and the inequality

$$F(m,n;p,2q) \ge \binom{m}{p} F(p,n;p,2q). \tag{10}$$

COROLLARY 1

$$F(m,n;p,2q) = \frac{1}{2}\binom{m}{p}\binom{n}{2q}\left(1 - \frac{C(q)\theta}{n^q}\right),$$

where $0 \le \theta = \theta(m,n;p,2q) \le 1$.

Proof is immediate from Theorem 1 and (4). ■

Let us write $f \sim g$ if $f/g \to 1$, $f \ll g$ (read "f is much less than g") if $f = o(g)$, and $f \asymp g$ if $f = O(g)$ and $g = O(f)$.

Corollary 1 implies in turn the following asymptotic result.

COROLLARY 2 *If* $n \to \infty$ *and* $q = O(1)$, *then*

$$F(m,n;p,2q) \sim \frac{1}{2}\binom{m}{p}\binom{n}{2q}.$$

From the proof of Lemma 2, one could obtain (5) with $C(q) = q(q^3 2^{2q+2}(2q)!)^q$ and thence slightly extend the zone $q = O(1)$ in Corollary 2 to $q < c\ln n/\ln\ln n$ for any $c \in (0, 1/2)$ and all sufficiently large n; somewhat tightening these estimates, it is possible to extend the zone for q further to $q \ll \sqrt{n}$; we have chosen not to give here details on that. Attracting some additional ideas instead, we extend the asymptotics of Corollary 2 to the still wider zone $q \ll n$, as stated in Theorem 2 below. On the other hand, Theorem 1 is not completely absorbed by Theorem 2, for the former provides the bound $O(n^{-q})$ on the relative error when q is bounded. Similar bounds could of course be obtained under the more general conditions of Theorem 2 as well, but that would overload this paper.

THEOREM 2 *If $n \to \infty$ and $q \ll n$, then*

$$F(m,n;\,p,2q) \sim \frac{1}{2}\binom{m}{p}\binom{n}{2q}.$$

The proof of Theorem 2 is preceded by another series of lemmata.
Introduce

$$\rho_j := \frac{a_j}{a_{j-1}},$$

where a_j are the same as in (7); note that ρ_j is defined only if $a_{j-1} \neq 0$, that is, if j is an integer satisfying the inequalities $\max(0, q+k-n/2) \leq j-1 \leq \min(k,q)$.

LEMMA 3 *If $n \to \infty$, $q \ll n$, and $k \ll n$, then ρ_j is defined and decreasing in j for all sufficiently large n and all $j = 1, \ldots, 1+\min(k,q)$.*

Proof is immediate upon looking at the expression

$$\rho_j = \frac{(2q-2j+1)(2q-2j+2)(k-j+1)}{(n-2q-2k+2j)(n-2q-2k+2j-1)j},$$

with the numerator decreasing and denominator increasing in j. ■

LEMMA 4 *If $n \to \infty$, $q \ll n$, $k \ll n$, and $\rho_1 \leq 1$, then*

$$g(n,k;q) \geq \frac{1}{2}\binom{n}{2q}$$

for all sufficiently large n.

Proof. By Lemma 3, $\rho_1 \leq 1$ implies $\rho_j \leq 1$ for all $j = 1, \ldots, 1+\min(k,q)$ and hence $a_{j-1} \geq a_j$ for all $j = 1, 2, \ldots$. This and (7) yield the Lemma. ■

LEMMA 5 *If $n \to \infty$, $q \ll n$, $k \ll n$, and $qk/n = O(1)$, then $\rho_1 \leq 1$ for all sufficiently large n.*

Proof is immediate from

$$\rho_1 = \frac{(2q-1)2qk}{(n-2q-2k+2)(n-2q-2k+1)} = O\left(\frac{qk}{n}\frac{q}{n}\right) = o(1). \quad ■$$

Introduce

$$b_i := \binom{k}{i}\binom{n-k}{2q-i}, \qquad i_* := \frac{(2q+1)(k+1)}{n+2},$$

$$\mu := \frac{1}{\binom{n}{2q}}\sum_i ib_i, \qquad \sigma := \left[\frac{1}{\binom{n}{2q}}\sum_i (i-\mu)^2 b_i\right]^{1/2},$$

$$r_i := \frac{b_i}{b_{i-1}};$$

note that r_i is defined only if $b_{i-1} \neq 0$, that is, if i is an integer satisfying the inequalities $\max(0, 2q+k-n) \leq i-1 \leq \min(k, 2q)$.

LEMMA 6 *If $n \to \infty$, $q \ll n$, $i_* \to \infty$, and $|i - i_*| \ll i_*$, then r_i is defined and converges to 1.*

Proof. Note that $i_* \to \infty$ implies $q \to \infty$ and $k \to \infty$, because $\max(k, q) \leq n/2$. Next, $i_* \leq \min(k/2, q+1/2)$ because $q \ll n$ and $k \leq n/2$; hence, the relations $|i - i_*| \ll i_*$, $q \ll n$, $k \leq n/2$, and $i_* \to \infty$ imply $\max(0, 2q+k-n) \leq i-1 \leq \min(k, 2q)$, so that r_i is defined. The convergence of r_i to 1 now follows from

$$r_i - 1 = -\frac{(n+2)(i-i_*)}{(n-2q-k+i)i},$$

taking again into account that $q \ll n$ and $k \leq n/2$. ■

LEMMA 7 *One has*

$$\text{(i)}\ \sum_i b_i = \binom{n}{2q}, \qquad \text{(ii)}\ \mu = \frac{2qk}{n}, \qquad \text{(iii)}\ \sigma^2 = \frac{2qk(n-k)(n-2q)}{n^2(n-1)}.$$

Proof. Identity (i) follows from the well-known Vandermonde convolution formula; see, e.g., (3) on page 8 of [2]. Next, (ii) is implied by (i) together with formula 6 on page 623 of [3]. [Alternatively, (i) and (ii) could be proved in a manner similar to that of the proof of (iii) below.] It remains to prove (iii). Toward that end, introduce

$$\mu_2 := \frac{1}{\binom{n}{2q}}\sum_i i(i-1)b_i.$$

Then $\binom{n}{2q}\mu_2$ is the coefficient of x^{2q} in the polynomial

$$\left[\sum_i i(i-1)\binom{k}{i}x^i\right]\left[\sum_j \binom{n-k}{j}x^j\right] = \left[x^2\frac{d^2}{dx^2}(1+x)^k\right](1+x)^{n-k}$$
$$= k(k-1)x^2(1+x)^{n-2},$$

whence

$$\mu_2 = k(k-1)\frac{2q(2q-1)}{n(n-1)}.$$

Using this, together with the identity $\sigma^2 = \mu_2 + \mu - \mu^2$ [easily following from (i)] and (ii), one has (iii). ■

Let us define the indicator function I of an assertion $\mathcal{A}$ by the formulae $\mathrm{I}\{\mathcal{A}\} = 1$ if $\mathcal{A}$ is true and $\mathrm{I}\{\mathcal{A}\} = 0$ if $\mathcal{A}$ is false.

The idea of the next lemma is to use the Chebyshev type inequality $\mathrm{I}\{x \geq y\} \leq x^2/y^2$ for any positive numbers x and y, as in Probability Theory.

LEMMA 8 *If $n \to \infty$, $q \ll n$, $qk/n \to \infty$, and a variable M is such that $\sigma \ll M$, then*

$$\sum_i b_i \mathrm{I}\{|i - i_*| > M\} \ll \binom{n}{2q}.$$

Proof. By Lemma 7 (iii), $\sigma^2 \asymp qk/n \to \infty$. Hence, $M \to \infty$ because $\sigma \to \infty$ and $\sigma \ll M$. Using now Lemma 7 (ii) and recalling the definition of i_*, one has

$$|i_* - \mu| = O\left(\frac{q+k+1}{n} + \frac{(2q+1)(k+1)}{n^2}\right) = O(1) \ll M.$$

Thus,

$$\begin{aligned} \sum_i b_i \mathrm{I}\{|i - i_*| > M\} &\leq \sum_i b_i \mathrm{I}\left\{|i - \mu| > \frac{M}{2}\right\} \\ &\leq \frac{4}{M^2} \sum_i b_i (i-\mu)^2 = \frac{4\sigma^2}{M^2}\binom{n}{2q} \ll \binom{n}{2q}, \end{aligned}$$

since $\sigma \ll M$. ■

LEMMA 9 *If $n \to \infty$, $q \ll n$, and $qk/n \to \infty$, then*

$$\sum_j b_{2j} \sim \frac{1}{2}\binom{n}{2q}.$$

Proof. The condition $qk/n \to \infty$ implies $k \to \infty$ and $q \to \infty$, because $\max(q,k) \leq n/2$. Recalling the definition of i_* and Lemma 7 (iii), one has $i_* \sim 2qk/n \asymp \sigma^2 \to \infty$; in particular, $\sigma \ll i_*$. This allows one to find varying with n, q, and k real numbers M_+ and M_- with the following properties: $M_+ > 0$, $M_- > 0$, $\sigma \ll M_+ \ll i_*$, $\sigma \ll M_- \ll i_*$, $i_* + M_+ =: 2j_+$ is an even integer, and $i_* - M_- =: 2j_- - 1$ is an odd integer; for instance, one can set $j_\pm = \lfloor (i_* \pm M)/2 \rfloor$ and then $M_+ = 2j_+ - i_*$ and $M_- = i_* - 2j_- + 1$, where

M varies so that $\sigma \ll M \ll i_*$. Now,

$$\sum_{j=j_-}^{j_+} b_{2j} = \sum_{j=j_-}^{j_+} b_{2j-1} r_{2j-1} \sim \sum_{j=j_-}^{j_+} b_{2j-1}$$

by Lemma 6. Hence, using the obvious relation

$$\sum_{j=j_-}^{j_+} b_{2j} + \sum_{j=j_-}^{j_+} b_{2j-1} = \sum_{i=i_*-M_-}^{i_*+M_+} b_i$$

together with Lemma 8, one obtains

$$\sum_{j=j_-}^{j_+} b_{2j} \sim \frac{1}{2} \sum_{i=i_*-M_-}^{i_*+M_+} b_i = \frac{1}{2} \sum_i b_i + o\left(\binom{n}{2q}\right).$$

Combining this with the other application of Lemma 8:

$$\sum_j b_{2j} = \sum_{j=j_-}^{j_+} b_{2j} + o\left(\binom{n}{2q}\right)$$

and recaling Lemma 7 (i), one completes the proof of the Lemma. ■

LEMMA 10 *If $n \to \infty$ and $q \ll n$, then for all $k = 0, \ldots, n$,*

$$g(n, k; q) \geq \frac{1}{2}\binom{n}{2q}(1 - o(1)).$$

Proof. It suffices to consider the following three cases.

Case 1: $qk/n \to \infty$. Then

$$g(n, k; q) = \sum_j b_{2j} \sim \frac{1}{2}\binom{n}{2q}$$

by Lemma 9.

Case 2.1: $qk/n = O(1)$ and $k \ll n$. Then, by Lemmata 5 and 4,

$$g(n, k; q) \geq \frac{1}{2}\binom{n}{2q}$$

for all sufficiently large n.

Case 2.2: $qk/n = O(1)$ and $n = O(k)$. Then $q = O(1)$, and so,

$$g(n, k; q) \geq \frac{1}{2}\binom{n}{2q}(1 - o(1))$$

in view of Corollary 2 (with $m = p$) and Lemma 1.

Thus, in all the three cases the statement of the Lemma is true. ∎

Proof of Theorem 2 is now immediate from Lemmata 10 and 1, and inequalities (4) and (10). ∎

Can the condition $q \ll n$, which is in fact the only condition of Theorem 2, be further relaxed? Computer-aided computations suggest that the answer to this question is no; nevertheless, the phenomenon of high concentration of the distribution of the number of the even submatrices seems to continue to hold if q is outside the $q \ll n$ zone but still far enough from n. However, consideration of that phenomenon would clearly need additional ideas and transcend the goals of this paper stated in the Abstract.

Acknowledgements

The problem studied in this paper has originally been posed by Alexander Sidorenko. Thanks are due to Anant Godbole for having communicated the problem to the author. Theorem 2 has been added to the original version of the paper to address interest expressed by one of the referees concerning the case of q growing with n.

References

1. D. de Caen, D. L. Kreher, and J. Wiseman, On constructive upper bounds for the Turán numbers $T(n, 2r + 1, 2r)$, *Congressus Numerantum*, Vol. 65 (1988) pp. 277-280.
2. J. Riordan, *Combinatorial Identities*, Wiley, New York (1968).
3. A. P. Prudnikov, Yu. A. Brychkov, and O. I. Marichev, *Integrals and Series. Volume 1*, Gordon and Breach Science Publishers, Amsterdam (1986).

Designs, Codes and Cryptography, 9, 95–104 (1996)
© 1996 Kluwer Academic Publishers, Boston. Manufactured in The Netherlands.

k-Arcs, Hyperovals, Partial Flocks and Flocks

J. A. THAS jat@cage.rug.ac.be
Department of Pure Mathematics and Computer Algebra, University of Ghent, Krijgslaan 281, B-9000 Gent, Belgium

Communicated by: V. Tonchev

Received November 29, 1994; Accepted May 19, 1995

Abstract. Some recent results on k-arcs and hyperovals of $PG(2, q)$, on partial flocks and flocks of quadratic cones of $PG(3, q)$, and on line spreads in $PG(3, q)$ are surveyed. Also, there is an appendix on how to use Veronese varieties as tools in proving theorems.

Keywords: k-arc, hyperoval, partial flock, line spread, Veronese variety

1. k-Arcs, Ovals and Hyperovals

A k *arc* in $PG(2, q)$ is a set of k points no three of which are collinear. An arc K is *complete* if it is not properly contained in a larger arc. Otherwise, if $K \cup \{x\}$ is an arc for some point x of $PG(2, q)$ we say that x *extends* K.

Let K be k-arc of $PG(2, q)$. Then clearly $k \leq q+2$. By Bose [2], for q odd, $k \leq q+1$. Further, any irreducible conic of $PG(2, q)$ is a $(q+1)$-arc. It can be shown that each $(q+1)$-arc K of $PG(2, q)$, q even, extends to a $(q+2)$-arc $K \cup \{x\}$ (see e.g. [10]); the point x, which is uniquely defined by K, is called the *nucleus, kernel* or *knot* of K. The $(q+1)$-arcs of $PG(2, q)$ are called *ovals*; the $(q+2)$-arcs of $PG(2, q)$, q even, are called *complete ovals* or *hyperovals*.

The following celebrated theorem is due to Segre [24].

THEOREM 1.1 *In $PG(2, q)$, q odd, every oval is an irreducible conic.*

Let $K \cup \{x\}$ be a hyperoval in $PG(2, q)$, q even, with K an irreducible conic. If $y \in K$, then $(K \backslash \{y\}) \cup \{x\} = K'$ is an oval of $PG(2, q)$. Clearly $|K \cap K'| = q$. So for $q > 4$ the oval K' cannot be an irreducible conic. Hence for $q \geq 8$, q even, the plane $PG(2, q)$ always contains ovals which are not irreducible conics. It is easy to show that for $q \in \{2, 4\}$ any oval of $PG(2, q)$ is an irreducible conic. By Segre [25, 26], each hyperoval of $PG(2, 8)$ is the union of a conic and its nucleus, and in $PG(2, 2^h)$ with $h = 5$ and $h \geq 7$ there exist hyperovals not containing a conic as a subset. Lunelli and Sce [16] have shown that in $PG(2, 16)$ there is a hyperoval which is not the union of a conic and its nucleus; a similar result for $PG(2, 64)$ was shown by Penttila and Pinneri [21] (they discovered two such hyperovals).

All hyperovals of $PG(2, q)$, $q \leq 32$, are known; for $q = 16$ this is due to Hall [9], for $q = 32$ to O'Keefe and Penttila [17].

Now we mention some important theorems on the extendability of k-arcs.

THEOREM 1.2 ([27]) *(a) If q is even and $k > (q+2)/2$ then K is contained in a unique complete arc of $PG(2,q)$.*
(b) If q is odd and $k > (2q+4)/3$ then K is contained in a unique complete arc of $PG(2,q)$.

THEOREM 1.3(a) ([27]) *For q even with*

$$k > q - \sqrt{q} + 1$$

any k-arc K is embedded in a hyperoval, which is unique except when $q = k = 2$.
(b) ([11]). Assume that $K > \lambda(q), q$ odd, where

$$\lambda(q) = \begin{cases} q - \frac{\sqrt{q}}{2} + 2 & for\ q = p^h \geq 23^2, q \neq 5^5,\ with p\ prime\ and\ p \geq 5, \\ \frac{44}{45}q + 2 & forq = p^h \leq 19^2\ with\ p\ prime\ and\ p \geq 5, \\ 3101 & for\ q = 5^5. \end{cases}$$

Then any k-arc K is embedded in a unique conic.
(c) ([30]). For q odd, any k-arc K with

$$k > q - \frac{\sqrt{q}}{4} + \frac{25}{16}$$

is embedded in a unique conic.
(d) ([35]). Every k-arc K of $PG(2,q), q = 2^{2m+1}, m \geq 1$, with

$$k > q - \sqrt{2q} + 2$$

is contained in a unique hyperoval.
(e) ([34, 35]). Assume that $k > \lambda(q), q$ odd, where

$$\lambda(q) = \begin{cases} \frac{44}{45}p + \frac{40}{45} & for\ q = p\ prime\ , \\ q - \frac{\sqrt{pq}}{4} + \frac{29}{16}p + 1 & for\ q = p^{2m+1}, p\ prime\ , m \geq 1. \end{cases}$$

Then any k-arc K is embedded in a unique conic.

Fisher, Hirschfeld and Thas [6], and independently Boros and Szönyi [1], construct complete $(q - \sqrt{q} + 1)$-arcs for q a square and $q > 4$; in fact these arcs were already constructed by Kestenband [15], but not recognized to be complete. So for q a square and $q \neq 4$ the bound of Segre is best possible. These $(q - \sqrt{q} + 1)$-arcs can be described as follows. Let G be a cyclic subgroup of $PGL(3,q)$ acting regularly on $PG(2,q)$, q square. Let G_1 be the subgroup of order $q - \sqrt{q} + 1$ of G. Then the orbits of G_1 are complete $(q - \sqrt{q} + 1)$-arcs when $q \geq 9$. Further, in the odd case the bounds in Theorem 1.3 certainly are not best possible; also, examples show that the bound $q - \sqrt{q} + 1$ does not work for q an odd square (in $PG(2,9)$ there exists a complete 8-arc, see [10]).

2. The Known Hyperovals of $PG(2, q), q = 2^h$

Let $D(k)$, with $k \in \mathbb{N}\backslash\{0\}$, be the pointset

$$\{(0,1,0),(0,0,1)\} \cup \{(1,t,t^k) \parallel t \in GF(q)\}.$$

Now we list all known hyperovals of $PG(2, q)$, q even.

(a) $D(2^m)$, with $(m, h) = 1$; these are due to Segre [25]. Note that $D(2)$ gives a conic union its nucleus.

(b) $D(6)$, with h odd; these are also due to Segre [26].

(c) Let h be odd, $h \geq 3$. Define two automorphisms

$$x \mapsto x^\sigma \text{ and } x \mapsto x^\gamma$$

of $GF(q)$ as follows :

$$\sigma = 2^{(h+1)/2}, \gamma = \begin{cases} 2^m & \text{if } h = 4m - 1 \\ 2^{3m+1} & \text{if } h = 4m + 1. \end{cases}$$

Then it was shown by Glynn [7] that $D(\sigma + \gamma)$ and $D(3\sigma + 4)$ are hyperovals.

(d) Let h be odd. Define $\delta : GF(q) \to GF(q)$ by

$$\delta : x \mapsto x^{1/6} + x^{1/2} + x^{5/6}.$$

Then Payne [19] has shown that

$$D(\delta) = \{(0,1,0),(0,0,1)\} \cup \{(1,t,t^\delta) \parallel t \in GF(q)\}$$

is a hyperoval of $PG(2, q)$.

(e) Next we describe the hyperovals of Cherowitzo [3, 4]. Let $h = 2s + 1$,

$$\begin{aligned} \sigma : GF(q) &\to GF(q), x \mapsto x^{2^{s+1}}, \\ \zeta : GF(q) &\to GF(q), x \mapsto x^\sigma + x^{\sigma+2} + x^{3\sigma+4}. \end{aligned}$$

Then

$$D(\zeta) = \{(0,1,0),(0,0,1)\} \cup \{(1,t,t^\zeta) \parallel t \in GF(q)\}$$

is a hyperoval ≥ 5. It does not belong to the previous classes for $h \geq 5$.

(f) Now we describe the Subiaco hyperovals [5]. Let $q = 2^e, e \geq 4$, and choose $\delta \in GF(q)$ for which both $\delta^2 + \delta + 1 \neq 0$ and $\text{tr}(1/\delta) = 1$ (so that $t^2 + \delta t + 1 \neq 0$ for all $t \in GF(q)$). Put

$$s(t) = \frac{\delta^2 t^4 + \delta^2(1+\delta+\delta^2)t^3 + \delta^2(1+\delta+\delta^2)t^2 + \delta^2 t}{(t^2+\delta t+1)^2} + t^{1/2}.$$

Then $\{(0,1,0),(0,0,1)\} \cup \{(1,t,s(t)) \parallel t \in GF(q)\}$ is a hyperoval in $PG(2,q)$. For any $e \equiv 2 \pmod 4$ there are exactly two Subiaco hyperovals; for any $e \not\equiv 2 \pmod 4$ there is just one Subiaco hyperoval.

For $q = 16$, the Lunelli-Sce hyperoval (see Section 1) is obtained as a special case.

For $q = 32$, the corresponding hyperoval of Payne is obtained as a special case.

For $q = 64$, the two hyperovals of Penttila-Pinneri (see Section 1) are obtained as a special case.

(g) Finally, O'Keefe and Penttila [17] discovered a new hyperoval in $PG(2,32)$, and Penttila and Royle [22, 23] discovered a new hyperoval in $PG(2,q)$ with $q = 64, 256, 1024, 4096$.

Note that the classes (a)-(f) sometimes overlap for small values of q, but they are distinct for large values.

3. Flocks and Partial Flocks of Quadratic Cones

Let K be a quadratic cone with vertex x in $PG(3,q)$. A partition of $K\backslash\{x\}$ into q disjoint conics is called a *flock* of K. If L is a line of $PG(3,q)$ having no points in common with K, then the q planes through L but not through x intersect K in the elements of a flock F. Such a flock is called *linear*.

Independently, Thas and Walker [36] (see also [31]) discovered that to each flock of K there corresponds a translation plane of order q^2 and dimension at most two over its kernel; in 1987 Thas proves that to each flock of K there corresponds a generalized quadrangle of order (q^2, q). Exploiting this relationship between flocks, translation planes and generalized quadrangles, several new infinite classes of each of these objects were discovered. Note that the flock F is linear if and only if the corresponding plane is Desarguesian resp. the generalized quadrangle is classical.

A *partial flock* of the cone K is a set of k disjoint conics of $K\backslash\{x\}$. It was proved by Payne and Thas [20] that any partial flock of size $q-1$ can be completed to a flock; this theorem has an interesting application to translation planes, see [14].

Let $X_0X_1 = X_2^2$ be the equation of K and let $\pi_i : x_iX_0 + z_iX_1 + y_iX_2 + X_3 = 0$ be a plane not containing the vertex (0,0,0,1) of K, $i = 1,2,\ldots,k$. Then the k conics $C_i = \pi_i \cap K$, $i = 1,2,\ldots,k$, form a partial flock of K if and only if
for q odd,

$$(y_i - y_j)^2 - 4(x_i - x_j)(z_i - z_j) \text{ is a nonsquare whenever } i \neq j,$$

for q even,

$$\operatorname{tr}((x_i + x_j)(z_i + z_j)(y_i + y_j)^{-2}) = 1 \text{ whenever } i \neq j.$$

For q even we can put $x_i = x_t, z_i = z_t, y_i = t$ with $t \in I \subset GF(q)$ and $|I| = k$, and without loss of generality we may assume that $0 \in I$ and $x_0 = z_0 = 0$.

4. Partial Flocks and k-Arcs

Let q be even.

Let $F = \{C_t \parallel t \in I\}$ be a partial flock of K, with $\pi_t : x_t X_0 + z_t X_1 + t X_2 + X_3 = 0, t \in I \subset GF(q), 0 \in I, x_0 = z_0 = 0$.

THEOREM 4.1 ([5], see also [29]) *In $PG(2, q)$ the pointset*

$$F(a_1, a_2) = \{(0, 0, 1), (0, 1, 0)\} \cup \{(1, a_1^2 x_t + a_1 a_2 t + a_2^2 z_t, t^2) \parallel t \in I\}$$

is a $(k+2)$-arc, for all $(a_1, a_2) \neq (0, 0)$. Conversely, if for all $(a_1, a_2) \neq (0, 0)$ the set $F(a_1, a_2)$ is a $(k+2)$-arc, then the corresponding planes π_t define a partial flock of the cone K. Such a set of $(k+2)$-arcs is called a herd of $(k+2)$-arcs.

COROLLARY *To any flock F of K there corresponds a herd of hyperovals in $PG(2, q)$.*

APPLICATION *The hyperovals of Payne and the Subiaco hyperovals were derived from flocks.*

Relying on Theorem 4.1 the following results on partial flocks were obtained.

THEOREM 4.2 ([29]) *Let F be a partial flock of size k of the quadratic cone K of $PG(3, q)$, q even. If*

$$k > q - \sqrt{q} - 1 \quad \text{for } q \text{ a square,}$$

or

$$k > q - \sqrt{2q} \quad \text{for } q \text{ a nonsquare,}$$

then F can be extended uniquely to a flock.

This theorem is a corollary of the following more general result.

THEOREM 4.3 ([29]) *Let F be a partial flock of size $k, k \geq q/2$, of the cone K. If all $(k+2)$-arcs of $PG(2, q)$ are extendable to a hyperoval, then F is extendable to a unique flock.*

We have the following interesting corollary on partial line spreads of $PG(3, q)$.

THEOREM 4.4 ([29]) *(a) Let P be a partial line spread of $PG(3, q)$, q even, which is the union of k reguli $R_1, R_2, \ldots, R_k, k \geq q/2$, which mutually intersect in the common line L. If all $(k+2)$-arcs of $PG(2, q)$ are extendable to a hyperoval, then P extends to a line spread S of $PG(3, q)$. This line spread S is the union of q reguli which mutually intersect in the line L. Also, S is the unique line spread containing P.*

(b) Let P be a partial line spread of $PG(3, q)$, q even, which is the union of k reguli $R_1, R_2, \ldots, R_k$, which mutually intersect in the common line L. If

$$k > q - \sqrt{q} - 1 \quad \text{for } q \text{ a square,}$$

or

$$k > q - \sqrt{2q} \quad \text{for } q \text{ a nonsquare,}$$

then P extends to a line spread S of $PG(3, q)$. This line spread S is the union of q reguli which mutually intersect in the line L. Also, S is the unique line spread containing P.

Using the relationship between flocks and herds of hyperovals, Storme and Thas [29] obtained short proofs of the following well known theorems on flocks, ovoids and translation generalized quadrangles.

THEOREM 4.5 ([13]) *If the translation generalized quadrangle S of order $(q^2, q), q$ even, arises from a flock, then it is the classical generalized quadrangle $H(3, q^2)$ arising from the non-singular Hermitian variety in $PG(3, q^2)$.*

THEOREM 4.6 ([31]) *If all planes of the conics of the flock F of the quadratic cone K of $PG(3, q), q$ even, contain a common point, then F is linear.*

Theorem 4.6 was deduced by Thas from the following equivalent result of Glynn.

THEOREM 4.7 ([8]) *If an ovoid O of $PG(3, q), q$ even, is the union of q conics which are mutually tangent at a common point, then O is an elliptic quadric.*

Remark. Glynn [8] used Theorem 4.7 to prove that any inversive plane of even order either is Miquelian, or of Suzuki-Tits type, or of Hering type I.1.

5. Collineations of Hyperovals

Let O be a hyperoval of $PG(2, q), q$ even. The brute force method to determine the elements α of $PGL(3, q)$ and $P\Gamma L(3, q)$ which fix O, is to take a general point x of O and to require that $x^\alpha \in O$. In many cases this implies long, technical and tedious calculations. But for several hyperovals calculations can be considerably simplified by considering O as an algebraic curve. This method was used by Thas, Payne and Gevaert [33] to determine the collineations of the hyperovals of Payne. Recently this method was applied by O'Keefe and Thas [18] to the "hyperovals" of Cherowitzo and to the Subiaco hyperovals. The following results were obtained.

THEOREM 5.1 ([18]) *The identity is the only element of $PGL(3, q)$ fixing the hyperoval of Cherowitzo (see 2(e)), $h \geq 7$, and fixing the point (0,0,1) (resp. (0,1,0)).*

THEOREM 5.2 ([18]) *The only elements of $PGL(3, q)$, $q = 2^e$ and $e \not\equiv 2$ (mod 4) with $e \geq 4$, fixing the Subiaco hyperoval O are 1 and γ with*

$$\gamma : \begin{pmatrix} x_0' \\ x_1' \\ x_2' \end{pmatrix} = \begin{pmatrix} 0 & 1 & 0 \\ 1 & 0 & 0 \\ 0 & 0 & 1 \end{pmatrix} \begin{pmatrix} x_0 \\ x_1 \\ x_2 \end{pmatrix}.$$

Sketch of the proof for e odd. For e odd we may take $\delta = 1$, so the Subiaco hyperoval is given by

$$O = \{(0,0,1),(0,1,0)\} \cup \{(1,t,s(t)) \,\|\, t \in GF(q)\},$$

with

$$s(t) = \frac{t^4+t^3+t^2+t}{(t^2+t+1)^2} + t^{1/2}.$$

Let $\sigma \in P\Gamma L(3,q)$ with $(x_0,x_1,x_2)^\sigma = (x_0^2,x_1^2,x_2^2)$. Then O^σ is the pointset of the algebraic curve $\mathcal{C}$ with equation

$$(X_2^2+X_0X_1)(X_0^2+X_0X_1+X_1^2)^4 + (X_0^2X_1^8+X_1^2X_0^8)+(X_0^4X_1^6+X_1^4X_0^6)=0. \quad (1)$$

Let γ be the algebraic closure of $GF(q)$. The curve over γ with equation (1) is denoted by $\hat{\mathcal{C}}$. The curve $\hat{\mathcal{C}}$ has a unique multiple point (0,0,1) of multiplicity 8 and the two linear factors of $X_0^2+X_0X_1+X_1^2$ (conjugate in a quadratic extension of $GF(q)$) define the tangents to $\hat{\mathcal{C}}$ at $(0,0,1)$. Further, $\hat{\mathcal{C}}$ is irreducible.

Let θ be an element of $PGL(3,q)$ fixing O. If $\hat{\mathcal{C}}^\theta \neq \hat{\mathcal{C}}$, then the sum of the intersection multiplicities at the common points of $\hat{\mathcal{C}}$ and $\hat{\mathcal{C}}^\theta$ is at least $q+16$. By the theorem of Bézout we have $q+16 \leq 100$. Hence $q \leq 64$.

If $q = 32$, then O is also a Payne hyperoval and by Thas, Payne and Gevaert [33] we have $\theta \in \{1,\gamma\}$.

So let $q \geq 128$. Then $\hat{\mathcal{C}}^\theta = \hat{\mathcal{C}}$. Hence θ fixes the multiple point (0,0,1) and the set of lines (in a quadratic extension of $GF(q)$) determined by the equation $X_0^2+X_0X_1+X_1^2=0$. Now it easily follows that $\theta \in \{1,\gamma\}$.

Appendix. Veronese Varieties as Tools in Proving Theorems

This section is taken from unpublished notes of Thas [32].

The following definitions and theorems on Veronese varieties are contained in Section 25.1 of [12].

The *Veronese variety* of all quadrics of $PG(n,F)$, $n \geq 1$, and F any commutative field, is the variety

$$\begin{aligned}\mathcal{V} = \{&(x_0^2,x_1^2,\ldots,x_n^2,x_0x_1,x_0x_2,\ldots,x_0x_n,x_1x_2,\ldots,x_1x_n,\ldots,x_{n-1}x_n)\,\|\\ &(x_0,x_1,\ldots,x_n) \text{ is a point of } PG(n,F)\}\end{aligned}$$

of $PG(N,F)$ with $N = n(n+3)/2$. Clearly $\mathcal{V}$ is a variety of dimension n. It is also called the *Veronesean* of quadrics of $PG(n,F)$, or simply the *quadric Veronesean* of $PG(n,F)$. It can be shown that the quadric Veronesean is absolutely irreducible and non-singular. Also, it generates $PG(N,F)$ and has order 2^n. Further $\mathcal{V}$ is the intersection of $(n+1)n^2/2$ quadrics.

Let $\zeta : PG(n,F) \to PG(N,F)$, with $N = n(n+3)/2$ and $n \geq 1$, be defined by $(x_0,x_1,\ldots,x_n) \mapsto (y_{00},y_{11},\ldots,y_{n-1,n})$, with $y_{ij} = x_ix_j$. It is an easy exercise to show that ζ is a bijection of $PG(n, F)$ onto the quadric Veronesean $\mathcal{V}$ of $PG(n, F)$. It then follows that the variety $\mathcal{V}$ is rational. Also, the quadrics of $PG(n, F)$ are mapped by ζ onto all hyperplane sections of $\mathcal{V}$.

Finally, the quadric Veronesean of $PG(n, F)$ only contains curves of even order; see Section 3 of Chapter VII in [28].

THEOREM A.1 *Each $(q+2)$-arc of $PG(2, q)$, q even with $q > 4$, has at least 6 points in common with an irreducible conic.*

Proof. Let K be a $(q+2)$-arc of $PG(2, q), q$ even and $q > 4$, and assume that no irreducible conic contains at least 6 points of K. The bijection ζ maps K onto a set $\tilde{K}$ of size $q+2$ on the quadric Veronesean $\mathcal{V}$ of $PG(2, q)$. No 6 points of $\tilde{K}$ are in a common hyperplane of the $PG(5, q)$ containing $\mathcal{V}$. So $\tilde{K}$ is a $(q+2)$-arc of $PG(5, q)$. This contradicts Theorem 27.6.11 of [12]. ■

COROLLARY *Each hyperoval of $PG(2, 8)$ is the union of a conic and its nucleus.*

Proof. Let K be a 10-arc of $PG(2, 8)$. By Theorem A.1 there is a conic C containing at least 6 points of K. Since $6 > (8+2)/2$, by Theorem 1.2(a) K coincides with $C \cup \{n\}$, where n is the nucleus of C. ■

THEOREM A.2 *Let every k-arc of $PG(5, q)$ with $q+1 \geq k > k_0 \geq 10$ be contained in a normal rational curve. Then in $PG(2, q)$ every k-arc with $q+1 \geq k > k_0$ has at least 6 points in common with an irreducible conic.*

Proof. Assume that every k-arc of $PG(5, q)$ with $q+1 \geq k > k_0 \geq 10$ is contained in a normal rational curve and consider a k-arc K with $q+1 \geq k > k_0$ of $PG(2, q)$. The bijection ζ maps K onto a set $\tilde{K}$ of size k on the quadric Veronesean $\mathcal{V}$ of $PG(2, q)$. Assume, by way of contradiction, that no irreducible conic contains 6 points of K. Then no 6 points of $\tilde{K}$ are in a common hyperplane of the $PG(5, q)$ containing $\mathcal{V}$. So $\tilde{K}$ is a k-arc of $PG(5, q)$. Since $q+1 \geq k > k_0 \geq 10$, the k-arc $\tilde{K}$ is contained in a normal rational curve $\mathcal{C}$ of $PG(5, q)$. The Veronesean $\mathcal{V}$ is the intersection of a set of quadrics $Q_1, Q_2, \ldots$. Since $|\tilde{K}| \geq 11$, the curve $\mathcal{C}$ (of order 5) and the quadric Q_i have at least 11 points in common. By the theorem of Bézout the curve $\mathcal{C}$ belongs to Q_i. Consequently $\mathcal{C}$ belongs to $\mathcal{V} = Q_1 \cap Q_2 \cap \cdots$. But $\mathcal{V}$ only contains curves of even order, a contradiction as $\mathcal{C}$ has order 5. We conclude that K has at least 6 points in common with an irreducible conic. ■

Example. By Theorem 27.7.23 of [12], for q even with $q \geq 8$ we can take $k_0 = q - \frac{1}{2}\sqrt{q} + \frac{17}{4}$. Hence in $PG(2, q), q = 2^h$ and $h \geq 3$, every k-arc K with $k > q - \frac{1}{2}\sqrt{q} + \frac{17}{4}$ has at least 6 points in common with an irreducible conic (for $k = q+2$ see Theorem A.1).

The previous theorems can be generalized as follows.

THEOREM A.3 *(a) Let K be a set of $q+2$ points of $PG(2,q)$ no four of which are on a line, with q even and $q \geq 8$. Then either there is an irreducible conic containing at least 6 points of K or there are two lines each intersecting K in 3 points and the common point of which does not belong to K.*

(b) Let every k-arc of $PG(5,q)$ with $q+1 \geq k > k_0 \geq 10$ be contained in a normal rational curve. Then for every set of k points of $PG(2,q)$, no four of which are on a line and with $q+1 \geq k > k_0$, we have the same conclusion as in (a).

Proof. (a) Let K be a set of $q+2$ points of $PG(2,q)$ no four of which are on a line, with q even and $q \geq 8$. Assume, by way of contradiction, that there is no conic (reducible or irreducible) containing at least 6 points of K. The bijection ζ maps K onto a set $\tilde{K}$ of size $q+2$ on the quadric Veronesean $\mathcal{V}$ of $PG(2,q)$. No 6 points of $\tilde{K}$ are in a common hyperplane of the $PG(5,q)$ containing $\mathcal{V}$. So $\tilde{K}$ is a $(q+2)$-arc of $PG(5,q)$. Then by Theorem 27.6.11 of [12] we have $q \leq 4$, a contradiction. Hence there is a conic containing at least 6 points of K, that is, either there is an irreducible conic containing at least 6 points of K or there are two lines each intersecting K in 3 points and the common point of which does not belong to K.

(b) Replace "irreducible conic" by "conic (irreducible or reducible)" in the proof of Theorem A.2. ■

Also quadric Veroneseans of $PG(n,q)$ can be used as tools in proving theorems. We illustrate this with one theorem.

THEOREM A.4 *Let every k-arc of $PG(9,q)$ with $q+1 \geq k > k_0 \geq 18$ be contained in a normal rational curve. Then in $PG(3,q)$ every k-arc with $q+1 \geq k > k_0$ has at least 10 points in common with an irreducible quadric.*

Proof. Similar to the proof of Theorem A.2. ■

Example. By Theorem 27.7.23 of [12], for q even with $q \geq 16$ we can take $k_0 = q - \frac{1}{2}\sqrt{q} + \frac{33}{4}$.

Remark. Theorem A.4 can be generalized by replacing "k-arc in $PG(3,q)$" by "set of size k in $PG(3,q)$" and "irreducible quadric" by "quadric".

Finally, also the Veronese variety of all hypersurfaces of order d of $PG(n,q)$ can be used as a tool for proving theorems on k-arcs of $PG(n,q)$ and more generally on pointsets of size k of $PG(n,q)$.

References

1. E. Boros, and T. Szönyi, On the sharpness of a theorem of B. Segre, *Combinatorica* Vol. 6 (1986) pp. 261–268.
2. R. C. Bose, Mathematical theory of the symmetrical factorial design, *Sankhyā* Vol. 8 (1947) pp. 107–166.
3. W. E. Cherowitzo, Hyperovals in Desarguesian planes of even order, *Ann. Discrete Math.* Vol. 37 (1986) pp. 87–94.
4. W. E. Cherowitzo, α-flocks and hyperovals, preprint.

5. W. E. Cherowitzo, T. Penttila, I. Pinneri, and G. F. Royle, Flocks and ovals, *Geom. Dedicata* Vol. 60 (1996) pp. 17–37.
6. J. C. Fisher, J. W. P. Hirschfeld, and J. A. Thas, Complete arcs in planes of square order, *Ann. Discrete Math.* Vol. 30 (1986) pp. 243–250.
7. D. G. Glynn, Two new sequences of ovals in finite Desarguesian planes of even order. In *Lecture Notes in Mathematics*, Vol. 1036, Springer, Berlin, pp. 217–229 (1983).
8. D. G. Glynn, The Hering classification for inversive planes of even order, *Simon Stevin* Vol. 58 (1984) pp. 319–353.
9. M. Hall, Ovals in the Desarguesian plane of order 16, *Ann. Mat. Pura Appl.* Vol. 102 (1975) pp. 159–176.
10. J. W. P. Hirschfeld, *Projective Geometry Over Finite Fields*, Oxford University Press, Oxford (1979).
11. J. W. P. Hirschfeld and G. Korchmáros, On the embedding of an arc into a conic in a finite plane, *Finite Fields Appl.*, to appear.
12. J. W. P. Hirschfeld and J. A. Thas, *General Galois Geometries*, Oxford University Press, Oxford (1991).
13. N. L. Johnson, Semifield flocks of quadratic cones, *Simon Stevin* Vol. 61 (1987) pp. 313–326.
14. N. L. Johnson, Translation planes admitting Baer groups and partial flocks of quadric sets, *Simon Stevin* Vol. 63 (1989) pp. 167–188.
15. B. Kestenband, Unital intersections in finite projective planes, *Geom. Dedicata* Vol. 11 (1981) pp. 107–117.
16. L. Lunelli and M. Sce, k-Archi completi nei piani proiettivi desarguesiani di rango 8 e 16. Technical Report, *Centro di Calcoli Numerici*, Politecnico di Milano (1958).
17. C. M. O'Keefe and T. Penttila, A new hyperoval in $PG(2, 32)$, *J. Geom.* Vol. 44 (1992) pp. 117–139.
18. C. M. O'Keefe and J. A. Thas, Collineations of Subiaco and Cherowitzo hyperovals, *Bull. Belg. Math. Soc. Simon Stevin* Vol. 3 (1996) pp. 177–192.
19. S. E. Payne, A new infinite family of generalized quadrangles, *Congress. Numer.* Vol. 49 (1985) pp. 115–128.
20. S. E. Payne and J. A. Thas, Conical flocks, partial flocks, derivation and generalized quadrangles, *Geom. Dedicata* Vol. 38 (1991) pp. 229–243.
21. T. Penttila and I. Pinneri, Irregular hyperovals in $PG(2, 64)$, *J. Geom.* Vol. 51 (1994) pp. 89–100.
22. T. Penttila and G. F. Royle, Classification of hyperovals in $PG(2, 32)$, *J. Geom.* Vol. 50 (1994) pp. 151–158.
23. T. Penttila and G. F. Royle, Personal communication.
24. B. Segre, Curve razionali normali e k-archi negli spazi finiti, *Ann. Mat. Pura Appl.* Vol. 39 (1955) pp. 357–379.
25. B. Segre, Sui k-archi nei piani finiti di caratteristica due, *Rev. Math. Pures Appl.* Vol. 2 (1957) pp. 289–300.
26. B. Segre, Ovali e curve σ nei piani di Galois di caratteristica due, *Atti Accad. Naz. Lincei Rend.* Vol. 32 (1962) pp. 785–790.
27. B. Segre, Introduction to Galois geometries (J. W. P. Hirschfeld, ed.), *Atti Accad. Naz. Lincei Mem.* Vol. 8 (1967) pp. 133–236.
28. J. G. Semple and L. Roth, *Introduction to Algebraic Geometry*, Oxford University Press, Oxford (1985).
29. L. Storme and J. A. Thas, k-Arcs and partial flocks, *Linear Algebra Appl.* Vol. 226 (1995) pp. 33–45.
30. J. A. Thas, Complete arcs and algebraic curves in $PG(2, q)$, *J. Algebra* Vol. 106 (1987) pp. 451–464.
31. J. A. Thas, Generalized quadrangles and flocks of cones, *European J. Combin.* Vol. 8 (1987) pp. 441–452.
32. J. A. Thas, Unpublished notes (1993).
33. J. A. Thas, S. E. Payne, and H. Gevaert, A family of ovals with few collineations, *European J. Combin.* Vol. 9 (1988) pp. 353–362.
34. J. F. Voloch, Arcs in projective planes over prime fields, *J. Geom.* Vol. 38 (1990) pp. 198–200.
35. J. F. Voloch, Complete arcs in Galois planes of non-square order. In *Advances in finite geometries and designs*, Oxford University Press, Oxford, pp. 401–406 (1991).
36. M. Walker, A class of translation planes, *Geom. Dedicata* Vol. 5 (1976) pp. 135–146.

Designs, Codes and Cryptography, 9, 105–114 (1996)
© 1996 Kluwer Academic Publishers, Boston. Manufactured in The Netherlands.

A Generalization of a Theorem of Dehon for Simple *t*-Designs

TRAN VAN TRUNG*
Institute for Experimental Mathematics, University of Essen, Ellernstrasse 29, 45326 Essen, Germany

Communicated by: V. Tonchev

Received December 9, 1994; Accepted February 2, 1995

Abstract. A generalization of a theorem of Dehon for simple t-designs is proved. Several new families of 4- and 5-designs having certain additional property regarding their block intersection numbers are obtained by applying this result.

Keywords: Simple t-design, mutually disjoint designs, permutation

1. Introduction

Dehon [2] proved that if there is a simple $t-(w,k,\lambda')$ design and a (not necessarily simple) $t-(v,w,\lambda)$ design such that ${\lambda'_0}^2(\lambda-1) < \binom{w}{k}$ then there exists a simple $t-(v,k,\lambda\lambda')$ design. This result is a generalization of the theorems of Tran van Trung [4] and S. S. Magliveras and T. E. Plambeck [3]. For more details about the connection between [2], [3] and [4] we refer to [6]. Our purpose in this paper is to present a generalization of the theorem of Dehon. The result is then applied to obtain new infinite families of 4- and 5-designs having certain additional property regarding their block intersection numbers.

For the reader's ease we quote some basic definitions and notations about t-designs adopted from [6]. A t-design, denoted by $t-(v,k,\lambda)$, is a pair $(X,\mathcal{B})$ where $\mathcal{B}$ is a collection of k-subsets (called blocks) of a v-set X (called points) such that every t-subset of X is in exactly λ blocks of $\mathcal{B}$. A t-design is said to have repeated blocks if two distinct blocks of $\mathcal{B}$ contain the same subset of X. A t-design is called simple if no blocks are repeated. A t-design is trivial if every k-subset of X is a block and occurs exactly m $(m \geq 1)$ times in $\mathcal{B}$. We are interested in constructing nontrivial simple t-designs.

Let $\mathcal{D}=(X,\mathcal{B})$ be a $t-(v,k,\lambda)$ design, let I be an i-subset of X with $0 \leq i \leq t$ and let λ_i be the number of blocks of $\mathcal{D}$ containing I. It is well-known that

$$\lambda_i = \lambda\binom{v-i}{t-i}\Big/\binom{k-i}{t-i},$$

where we have $\lambda=\lambda_t$ and λ_0 is the total number of blocks.

* The author acknowledges the support of SFB Forschungsschwerpunkt-Geometrie Grant, University of Heidelberg.

Let $(X, \mathcal{B})$ be a $t-(v, k, \lambda)$ design and let $(X', \mathcal{B}')$ be a $t-(v', k', \lambda')$ design. $(X, \mathcal{B})$ and $(X', \mathcal{B}')$ are said to be disjoint if $\mathcal{B} \cap \mathcal{B}' = \emptyset$. Let $(X_i, \mathcal{B}_i)$ be a $t-(v_i, k_i, \lambda^{(i)})$ design for $i = 1, \ldots, m$; we say that $(X_1, \mathcal{B}_1), \ldots, (X_m, \mathcal{B}_m)$ are mutually disjoint if $(X_i, \mathcal{B}_i)$ and $(X_j, \mathcal{B}_j)$ are disjoint for $i, j = 1, \ldots, m,\ i \neq j$.

If X be a set, then $Sym(X)$ denotes the symmetric group of X.

2. The Generalization

We begin with two technical lemmas.

LEMMA 2.1 *Let $\mathcal{D} = (X, \mathcal{B})$ be a $t-(v, k, \lambda)$ simple design, let $\Delta \subset X$ be a d-subset of X and let h be an integer such that $k - h \geq t$. Then the number of blocks B of $\mathcal{D}$ with $|B \cap \Delta| = k - h$ is at most*

$$\lambda \frac{d(d-1)\ldots(d-t+1)}{(k-h)(k-h-1)\ldots(k-h-t+1)}.$$

Proof. The result is obtained by counting in two ways the number of pairs (T, B) such that $T \subset B \in \mathcal{B}$, $|T| = t$ and $|B \cap \Delta| = k - h$.

(α) For any $T \subseteq X$, $|T| = t$ there are at most λ blocks B with $T \subset B \cap \Delta$ and $|B \cap \Delta| = k - h$. Hence there are at most $\lambda\binom{d}{t}$ pairs (T, B) with the required property.

(β) Let ϕ be the number of blocks B such that $|B \cap \Delta| = k - h$. For any block B with $|B \cap \Delta| = k - h$ there are $\binom{k-h}{t}$ t-subsets T contained in $B \cap \Delta$. Thus there are $\phi\binom{k-h}{t}$ pairs (T, B).

Comparing (α) and (β) gives

$$\lambda\binom{d}{t} \geq \phi\binom{k-h}{t},$$

hence

$$\phi \leq \frac{\lambda\binom{d}{t}}{\binom{k-h}{t}} = \lambda \frac{d(d-1)\ldots(d-t+1)}{(k-h)(k-h-1)\ldots(k-h-t+1)}. \qquad \blacksquare$$

The next lemma is due to Dehon [2].

LEMMA 2.2 *Let $\mathcal{D} = (X, \mathcal{B})$ be a $t-(v, w, \lambda)$ design and let $\mathcal{B} = \{B_1, B_2, \ldots, B_{\lambda_0}\}$. For every $1 \leq i, j \leq \lambda_0$, let $m_{ij} = |B_i \cap B_j|$. Then*

$$\sum_{\substack{j=1 \\ j \neq i}}^{\lambda_0} m_{ij}(m_{ij}-1)\ldots(m_{ij}-t+1) = (\lambda-1)w(w-1)\ldots(w-t+1)$$

for $i = 1, 2, \ldots, \lambda_0$.

Proof. Let $B_i \in \mathcal{B}$. Count in two ways the number of pairs (T, B_j) where T is a t-subset of B_i, $j \neq i$ and $T \subset B_j$.

(α) For any t-subset $T \subset B_i$ there are $(\lambda - 1)$ blocks B_j, $j \neq i$, containing T. Hence there are $\binom{w}{t}(\lambda - 1)$ pairs (T, B_j).

(β) Set $M_{ij} = B_i \cap B_j$, $i \neq j$. Any M_{ij} (corresponding to a block B_j) contains $\binom{m_{ij}}{t}$ t-subsets T. Thus the number of pairs (T, B_j) is $\sum_{\substack{j=1\\ j\neq i}}^{\lambda_0} \binom{m_{ij}}{t}$.

Equating (α) and (β) we have

$$\sum_{\substack{j=1\\ j\neq i}}^{\lambda_0} \binom{m_{ij}}{t} = (\lambda - 1)\binom{w}{t}$$

or equivalently

$$\sum_{\substack{j=1\\ j\neq i}}^{\lambda_0} m_{ij}(m_{ij} - 1)\ldots(m_{ij} - t + 1) = (\lambda - 1)w(w - 1)\ldots(w - t + 1). \qquad \blacksquare$$

For simplicity, we say that two designs $\mathcal{D} = (X, \mathcal{B})$ and $\mathcal{D}' = (X', \mathcal{B}')$ with parameters $t - (v, k, \lambda)$ and $t - (v', k, \lambda')$, respectively, satisfy the P_{k-h} property if $|B \cap B'| < k - h$ for any two blocks $B \in \mathcal{D}$ and $B' \in \mathcal{D}'$.

The following theorem is our main result.

THEOREM 2.3 *Let $\mathcal{D} = (X, \mathcal{B})$ be a $t - (v, w, \lambda)$ design, not necessarily simple. Let $\mathcal{D}' = (X', \mathcal{B}')$ be a simple $t - (w, k, \lambda')$ design with $|B' \cap C'| < k - h$ for any two distinct blocks $B', C' \in \mathcal{B}'$ and $0 \leq h \leq k$. If*

$$\lambda_0'^2(\lambda - 1)\binom{k}{h}\binom{k}{t}\sum_{i=0}^{h} \frac{\binom{k-i}{h-i}}{\binom{k-i}{t}} < \binom{w}{k-h},$$

then there exists a simple $t - (v, k, \lambda\lambda')$ design in which any two distinct blocks meet in less than $k - h$ points.

Proof. Let $\mathcal{B} = \{B_1, B_2, \ldots, B_{\lambda_0}\}$. On each B_i, considered as a point set, construct a design $\mathcal{D}_i = (B_i, \mathcal{B}_i)$ isomorphic to $\mathcal{D}'$. If we do it for all the blocks $B_1, B_2, \ldots, B_{\lambda_0}$, then we get a $t - (v, k, \lambda\lambda')$ design, which is in general not simple, because $\mathcal{D}$ need not be simple. The idea of the proof is the following:

Assume that a set of $n-1$ mutually disjoint designs $\mathcal{D}'_1, \mathcal{D}'_2, \ldots, \mathcal{D}'_{n-1}$ have been found, in which any pair $(\mathcal{D}'_i, \mathcal{D}'_j)$, $i \neq j$, $i, j = 1, \ldots, n-1$, has the P_{k-h} property. We find a condition for which a permutation $\sigma_n \in Sym(B_n)$ exists such that any block of $\mathcal{D}'_n = (B_n, \mathcal{B}_n^{\sigma_n})$, where $\mathcal{B}_n^{\sigma_n} = \{C^{\sigma_n}/C \in \mathcal{B}_n\}$, intersects any block of $\mathcal{D}'_j$, $j = 1, \ldots, n-1$, in less than $k - h$ points; i.e. the pair $(\mathcal{D}'_n, \mathcal{D}'_j)$, $j = 1, \ldots, n-1$, has the P_{k-h} property.

We start with $\mathcal{D}'_1 = (B_1, \mathcal{B}_1)$. Assume that for some $n > 1$ a set of $n - 1$ designs $\mathcal{D}'_i = (B_i, \mathcal{B}_i^{\sigma_i})$ exists for $i = 1, \ldots, n-1$, where $\sigma_i \in Sym(B_i)$, such that any pair

$(\mathcal{D}'_i, \mathcal{D}'_j)$ has the P_{k-h} property for $i \neq j,\ i, j = 1, \ldots, n-1$. Note that for $\mathcal{D}'_1$ we choose σ_1 as the identity of $Sym(B_1)$.

Set

$$m_{ij} = |B_i \cap B_j|, \quad i, j = 1, \ldots, \lambda_0.$$

Let

$$r_{jn}^{(k-i)} = |\{C \in \mathcal{B}_j^{\sigma_j} / \, |C \cap B_n| = k-i\}|, \quad \text{for } j = 1, \ldots, n-1,$$

and

$$r_{jn}^{(k-i)} = |\{C \in \mathcal{B}_j / \, |C \cap B_n| = k-i\}|, \quad \text{for } j = n+1, \ldots, \lambda_0.$$

For each $j < n$ we find the number of permutations $\varphi \in Sym(B_n)$ such that $\mathcal{D}'_j$ and $(B_n, \mathcal{B}_n^{\varphi})$ do not have the P_{k-h} property. Let C be a block of $\mathcal{D}'_j$ with $|C \cap B_n| = k-i$. For each of the $r_{jn}^{(k-i)}$ such blocks C, the number of permutations $\varphi \in Sym(B_n)$ sending any $(k-h)$-subset Δ of $C \cap B_n$ to any $(k-h)$-subset Γ of a block of $\mathcal{D}_n$ is:

$$(k-h)!\binom{k-i}{h-i}(w-k+h)!\binom{k}{h}.$$

This can be seen as follows: take a pair (Δ, Γ); the number of permutations $\varphi \in Sym(B_n)$ with $\Delta^{\varphi} = \Gamma$ is $(k-h)!(w-k+h)!$. As there are $\binom{k-i}{h-i}$ $(k-h)$-subsets Δ in $C \cap B_n$ and $\binom{k}{h}$ $(k-h)$- subsets Γ in a block of $\mathcal{D}_n$, the claim follows.

Further, since there are λ'_0 blocks in $\mathcal{D}_n$, the number of permutations $\varphi \in Sym(B_n)$ such that $\mathcal{D}'_j$ and $(B_n, \mathcal{B}_n^{\varphi})$ do not have the property P_{k-h} is:

$$\sum_{i=0}^{h} r_{jn}^{(k-i)}(k-h)!(w-k+h)!\binom{k-i}{h-i}\binom{k}{h}\lambda'_0.$$

Therefore, if

$$\sum_{j=1}^{n-1}\sum_{i=0}^{h} r_{jn}^{(k-i)}(k-h)!(w-k+h)!\binom{k-i}{h-i}\binom{k}{h}\lambda'_0 < w!, \tag{1}$$

then there is at least one permutation $\sigma_n \in Sym(B_n)$ such that $\mathcal{D}'_n = (B_n, \mathcal{B}_n^{\sigma_n})$ and $\mathcal{D}'_j$, $j = 1, \ldots, n-1$, have the P_{k-h} property. Applying Lemma 2.1 to $\mathcal{D}'_j$ for $j = 1, \ldots, n-1$ and to $\mathcal{D}_j$ for $j = n+1, \ldots, \lambda_0$ we obtain

$$\sum_{j=1}^{n-1} \lambda'_0(k-h)!(w-k+h)!\binom{k}{h}\sum_{i=0}^{h} r_{jn}^{(k-i)}\binom{k-i}{h-i}$$

$$\leq \lambda'_0(k-h)!(w-k+h)!\binom{k}{h}\sum_{\substack{j=1\\ j\neq n}}^{\lambda_0}\sum_{i=0}^{h} r_{jn}^{(k-i)}\binom{k-i}{h-i}$$

$$\leq \lambda'_0(k-h)!(w-k+h)!\binom{k}{h}\sum_{i=0}^{h}\binom{k-i}{h-i}\frac{1}{\binom{k-i}{t}t!}\sum_{\substack{j=1\\ j\neq n}}^{\lambda_0} \lambda' m_{in}(m_{in}-1)\ldots(m_{in}-t+1)$$

Using Lemma 2.2 for $\mathcal{D}$ we see that the left side of (1) is upper-bounded by

$$\lambda_0'(k-h)!(w-k+h)!\binom{k}{h}\lambda'(\lambda-1)w(w-1)\dots(w-t+1)\sum_{i=0}^{h}\binom{k-i}{h-i}\frac{1}{\binom{k-i}{t}t!}$$

$$=\lambda_0'^2(\lambda-1)\binom{k}{h}\binom{k}{t}(k-h)!(w-k+h)!\sum_{i=0}^{h}\binom{k-i}{h-i}\frac{1}{\binom{k-i}{t}}.$$

It follows that if

$$\lambda_0'^2(\lambda-1)\binom{k}{h}\binom{k}{t}(k-h)!(w-k+h)!\sum_{i=0}^{h}\binom{k-i}{h-i}\frac{1}{\binom{k-i}{t}} < w!,$$

or equivalently

$$\lambda_0'^2(\lambda-1)\binom{k}{h}\binom{k}{t}\sum_{i=0}^{h}\frac{\binom{k-i}{h-i}}{\binom{k-i}{t}} < \binom{w}{k-h},$$

then there exists a permutation $\sigma_n \in Sym(B_n)$ such that $\mathcal{D}'_n = (B_n, \mathcal{B}_n^{\sigma_n})$ and $\mathcal{D}'_j, j = 1,\dots,n-1$, have the P_{k-h} property.

This proves the assertion. ■

Note that Theorem 1 of Dehon [2] is a special case of Theorem 2.3. In fact, taking $h=0$ the inequality in Theorem 2.3 becomes $\lambda_0'^2(\lambda-1) < \binom{w}{k}$, which is the condition in the Dehon's paper.

Moreover, for $h=1$ the inequality of Theorem 2.3 becomes

$$\lambda_0'^2(\lambda-1)k^2\frac{(k-t+1)}{(k-t)} < \binom{w}{k-1}. \tag{2}$$

3. Some Applications

In this section we show some applications of Theorem 2.3. For our purpose, we begin at first with the following remark concerning the designs obtained from a given design by enlarging its blocks.

Remark.

(*) If $\mathcal{D}=(X,\mathcal{B})$ is a $t-(v,k,\lambda)$ design then $\mathcal{D}^+=(X,\mathcal{B}^+)$ is a

$$t-(v,k+1,\lambda(v-k)\frac{k+1}{k-t+1})$$

design, where $\mathcal{B}^+=\{B\cup\{x\}/B\in\mathcal{B}, x\in X\setminus B\}$.

(**) If we use (*) for $\mathcal{D}^+ = (X, \mathcal{B}^+)$, then $\mathcal{D}^{++} = (X, \mathcal{B}^{++})$ is a

$$t - (v, k+2, \lambda(v-k)(v-k-1)\frac{(k+1)(k+2)}{(k-t+1)(k-t+2)})$$

design, where $\mathcal{B}^{++} = \{B \cup \{x, y\} / B \in \mathcal{B}, x, y \in X \setminus B,\ x \neq y\}$.

Of course, $\mathcal{D}^+$ and $\mathcal{D}^{++}$ are not simple in general. $\mathcal{D}^+$ is simple if and only if $|B \cap B'| < k-1$ for any pair (B, B') of blocks of $\mathcal{D}$.

Now the complement of the Alltop's 4-design [1]

$$4 - (2^{n+1}+1, 2^n, (2^n-3)(2^{n-1}-1))$$

is a

$$4 - (2^{n+1}+1, 2^n+1, (2^n+1)(2^{n-1}-1))$$

design $\mathcal{D}$, for $n \geq 3$. Using (*) $\mathcal{D}^+$ is a

$$4 - (2^{n+1}+1, 2^n+2, 2^n(2^n+1)(2^{n-1}+1))$$

design. Moreover, for $n \geq 6$ there exists a

$$4 - (2^n+2, 2^{n-1}+1, (2^n-2)(2^{n-2}-1))$$

design $\mathcal{D}'$ with $|B \cap B'| < 2^{n-1}$ for every pair (B, B') of blocks in $\mathcal{D}'$ [6].

Applying Theorem 2.3 to $\mathcal{D}^+$ and $\mathcal{D}'$ for $h = 1$ we see that the condition (2) is satisfied for $n \geq 6$. Thus we have the following theorem.

THEOREM 3.1 *For $n \geq 6$ there exists a*

$$4 - (2^{n+1}+1, 2^{n-1}+1, 2^n(2^n+1)(2^{n-1}+1)(2^{n-2}-1)(2^n-2))$$

design in which any two distinct blocks meet in less than 2^{n-1} points.

In the following we describe a procedure to show how Theorem 2.3 can be used recursively. This process is analogous to that for the Theorem 3 in [2].

Let

$$\mathcal{D}_{n,1} := \mathcal{D}^+ = 4 - (2^{n+1}+1, 2^n+2, 2^n(2^n+1)(2^{n-1}+1)),$$

and

$$\mathcal{D}'_n := \mathcal{D}' = 4 - (2^n+2, 2^{n-1}+1, (2^n-2)(2^{n-2}-1))$$

be the designs given in the proof of Theorem 3.1 above. Note that any two distinct blocks of $\mathcal{D}'_n$ intersect in less than 2^{n-1} points.

Denote by

$$\mathcal{C}_{n,1} := 4 - (2^{n+1} + 1, 2^{n-1} + 1, \lambda^{\mathcal{D}'_n} \lambda^{\mathcal{D}_{n,1}}), \quad n \geq 6,$$

i.e the design in Theorem 3.1, where $\lambda^{\mathcal{D}'_n}$ is the value λ of $\mathcal{D}'_n$ and $\lambda^{\mathcal{D}_{n,1}}$ that of $\mathcal{D}_{n,1}$. Thus $\mathcal{C}_{n+1,1}$ is a

$$4 - (2^{n+2} + 1, 2^n + 1, \lambda^{\mathcal{D}'_{n+1}} \lambda^{\mathcal{D}_{n+1,1}})$$

design. Using (*) for $\mathcal{C}_{n+1,1}$ we obtain a design

$$\mathcal{D}_{n,2} := 4 - \left(2^{n+2} + 1, 2^n + 2, \lambda^{\mathcal{D}'_{n+1}} \lambda^{\mathcal{D}_{n+1,1}} \frac{(2^{n+2} - 2^n)(2^n + 2)}{(2^n - 2)}\right)$$

Now applying Theorem 2.3 to $\mathcal{D}_{n,2}$ and $\mathcal{D}'_n$ for $h = 1$ we have a design

$$\mathcal{C}_{n,2} := 4 - \left(2^{n+2} + 1, 2^{n-1} + 1, \lambda^{\mathcal{D}'_n} \lambda^{\mathcal{D}'_{n+1}} \lambda^{\mathcal{D}_{n+1,1}} \frac{(2^{n+2} - 2^n)(2^n + 2)}{(2^n - 2)}\right)$$

In this way, using Theorem 2.3 recursively we can prove the existence of a family of designs $\mathcal{C}_{n,1}, \mathcal{C}_{n,2}, \ldots, \mathcal{C}_{n,d}$ of block size $2^{n-1} + 1$, in which any two blocks intersect in less than 2^{n-1} points. This process may be illustrated as follows:

$$\left.\begin{array}{c}\mathcal{D}_{n,1}\\ \mathcal{D}'_n\end{array}\right\} \mathcal{C}_{n,1} \longrightarrow \mathcal{C}_{n+1,1} \nearrow \left.\begin{array}{c}\mathcal{D}_{n,2}\\ \mathcal{D}'_n\end{array}\right\} \mathcal{C}_{n,2} \longrightarrow \mathcal{C}_{n+1,2} \nearrow \left.\begin{array}{c}\mathcal{D}_{n,3}\\ \mathcal{D}'_n\end{array}\right\} \mathcal{C}_{n,3}$$

$$\cdots \mathcal{C}_{n,d-1} \longrightarrow \mathcal{C}_{n+1,d-1} \nearrow \left.\begin{array}{c}\mathcal{D}_{n,d}\\ \mathcal{D}'_n\end{array}\right\} \mathcal{C}_{n,d}.$$

Diagram 1.

Thus we have the following theorem.

THEOREM 3.2 *For $n \geq 6$ there exists a function $f(n) > 0$ such that for every $d \leq f(n)$ there exists a*

$$4 - (2^{n+d} + 1, 2^{n-1} + 1, \lambda)$$

design with

$$\lambda = (2^{n+d-3} - 1)(2^{n+d-1} - 2)2^{n+d-1}(2^{n+d-1} + 1)(2^{n+d-2} + 1) \times \prod_{i=1}^{d-1}(2^{n+i-3} - 1)(2^{n+d} - 2^{n+i-1})(2^{n+i-1} + 2),$$

in which any two distinct blocks have less than 2^{n-1} points in common.

The inequality (2) for the design in Theorem 3.2 becomes

$$(\lambda^{\mathcal{D}_{n,d}} - 1)2^4(2^{2n} - 1)^2(2^{n-1} + 1)^2(2^{n-1} - 2)/(2^{n-1} - 3) < \binom{2^n + 2}{2^{n-1}}$$

or equivalently

$$(\lambda^{\mathcal{D}_{n,d}} - 1)\frac{2^4(2^{2n} - 1)^2(2^{n-1} + 1)^2(2^{n-1} - 2)}{(2^{n-1} - 3)}[2^{n-1}!]$$
$$\leq (2^n + 2)(2^n + 1)\dots(2^{n-1} + 3). \qquad (3)$$

The function $f(n)$ is derived from (3). It is not possible to have an explicit expression for $f(n)$. But a rough lower bound for $f(n)$, namely $f(n) \geq 2^{(n-1)/2}/\sqrt{3}$, for large n, can be seen as follows: Here

$$\lambda^{\mathcal{D}_{n,d}} = (2^{n+d-3} - 1)(2^{n+d-1} - 2)2^{n+d-1}(2^{n+d-1} + 1)(2^{n+d-2} + 1)$$
$$\times (2^{n+d} - 2^n)\frac{(2^n + 2)}{(2^n - 2)}\prod_{i=1}^{d-2}(2^{n+i-2} - 1)(2^{n+d} - 2^{n+i})(2^{n+i} + 2).$$

Put $A = 2^4(2^{2n} - 1)^2(2^{n-1} + 1)^2(2^{n-1} - 2)/(2^{n-1} - 3)$. Then the left side of (3) is lower than $\lambda^{\mathcal{D}_{n,d}}.A.2^{n-1}!$.

Clearly,

$$A < 2^{6n+5} \text{ and } \lambda^{\mathcal{D}_{n,d}} < 2^{6n+6d-4}\prod_{i=1}^{d-2}2^{3n-1+2i+d} = 2^{3nd+2d^2}.$$

Using Stirling's approximation we have

$$2^{n-1}! \approx \sqrt{2\pi}(2^{n-1})^{2^{n-1}+\frac{1}{2}}e^{-2^{n-1}} < 2^{2^{n-1}(n-2)+n+1}.$$

Thus the left side of (3) is lower than

$$2^{6n+5}2^{3nd+2d^2}2^{2^{n-1}(n-2)+n+1} = 2^{7n+6+3nd+2d^2+2^{n-1}(n-2)}.$$

On the other hand, the right side of (3) can be seen as the product of 2^{n-1} factors greater than 2^{n-1}; it is thus greater than $2^{2^{n-1}(n-1)}$. Hence the condition (3) is satisfied if

$$2^{7n+6+3nd+2d^2+2^{n-1}(n-2)} < 2^{2^{n-1}(n-1)}$$

or equivalently

$$7n + 6 + 3nd + 2d^2 < 2^{n-1}.$$

For large values of n, the left side of this inequality is lower than $3d^2$. Hence (3) is satisfied if $3d^2 \leq 2^{n-1}$ or $d \leq 2^{(n-1)/2}/\sqrt{3}$. In other words, $f(n) \geq 2^{(n-1)/2}/\sqrt{3}$ for large values of n.

We are now in a position to show the existence of some new infinite families of 5-designs.

There exists a

$$5-(2^{n+1}+4, 2^n+2, (2^{n+1}-1)(2^{n+1}-2)(2^{n-1}-1))$$

design, for $n \geq 4$, which we denote by $\mathcal{E}_{n+1,0}$ (see [5]). Using (**) we obtain from $\mathcal{E}_{n+1,0}$ a design $\mathcal{F}_{n,1}$. So, $\mathcal{F}_{n,1}$ is a $5-(2^{n+1}+4, 2^n+4, \lambda^{\mathcal{F}_{n,1}})$ design with

$$\lambda^{\mathcal{F}_{n,1}} = (2^{n+1}-1)(2^{n+1}-2^n+2)(2^{n+1}-2^n+1)(2^n+3)(2^n+4).$$

Let $\mathcal{F}'_n$ be a

$$5-(2^n+4, 2^{n-1}+2, (2^n-1)(2^n-2)(2^{n-1}-2)), \quad n \geq 6$$

design in which any two blocks intersect in less than $2^{n-1}+1$ points (see Theorem 6 [6]). Using Theorem 2.3 for $\mathcal{F}_{n,1}$ and $\mathcal{F}'_n$ with $h=1$ we have the following result.

THEOREM 3.3 *For $n \geq 6$ there exists a*

$$5-(2^{n+1}+4, 2^{n-1}+2, \lambda)$$

design with

$$\begin{aligned}\lambda = {} & (2^{n-1}-2)(2^{n+1}-1)(2^{n+1}-2)(2^{n-1}-1)(2^{n+1}-2^n+2) \\ & \times (2^{n+1}-2^n+1)(2^n+3)(2^n+4),\end{aligned}$$

in which any two distinct blocks meet in less than $2^{n-1}+1$ points.

Let $\mathcal{E}_{n,1}$ denote the design in Theorem 3.3. Using Theorem 2.3 for $h=1$ and proceeding by the same way as that of Theorem 3.2 we can prove the existence of a family of 5-designs $\mathcal{E}_{n,2}, \mathcal{E}_{n,3}, \ldots, \mathcal{E}_{n,d}$. This procedure is depicted in the diagram below:

$$\mathcal{E}_{n,1} \longrightarrow \mathcal{E}_{n+1,1} \nearrow \left.\begin{matrix}\mathcal{F}_{n,2}\\ \mathcal{F}'_n\end{matrix}\right\} \mathcal{E}_{n,2} \longrightarrow \mathcal{E}_{n+1,2} \nearrow \left.\begin{matrix}\mathcal{F}_{n,3}\\ \mathcal{F}'_n\end{matrix}\right\} \mathcal{E}_{n,3} \longrightarrow \mathcal{E}_{n+1,3} \nearrow \left.\begin{matrix}\mathcal{F}_{n,4}\\ \mathcal{F}'_n\end{matrix}\right\} \mathcal{E}_{n,4}$$

$$\cdots \mathcal{E}_{n,d-1} \longrightarrow \mathcal{E}_{n+1,d-1} \nearrow \left.\begin{matrix}\mathcal{F}_{n,d}\\ \mathcal{F}'_n\end{matrix}\right\} \mathcal{E}_{n,d}.$$

Diagram 2.

In Diagram 2, the design $\mathcal{F}_{n,i}$ is obtained from $\mathcal{E}_{n+1,i-1}$ by (**). It is clear that Theorem 2.3 can be used for d steps so far as the hypothesis is satisfied. The following theorem can be therefore considered as a generalization of Theorem 3.3.

THEOREM 3.4 *For $n \geq 7$ there exists a function $f(n) > 0$ such that for every $d \leq f(n)$ there exists a*

$$5-(2^{n+d}+4, 2^{n-1}+2, \lambda)$$

design with

$$\lambda = (2^{n+d}-1)(2^{n+d}-2)(2^{n+d-2}-1) \times \prod_{i=0}^{d-1}(2^{n+i-1}-2)(2^{n+d}-2^{n+i}+2)(2^{n+d}-2^{n+i}+1)(2^{n+i}+3)(2^{n+i}+4),$$

in which any two distinct blocks have less than $2^{n-1}+1$ *points in common.*

After some computation we can see that $f(n) \geq 2^{(n-3)/2}$ for large values of n in this case.

Note that each of the families of 4- or 5-designs in Theorem 3.1, 3.2, 3.3 and 3.4 also produces another new familly of simple 4- or 5-designs by the process of enlarging blocks given by $(*)$. Finally, new infinite families of 4- or 5-designs can be obtained as well by using Theorem 2.3 with $h = 1$ for appropriate 4- or 5-designs in [4], [5] and [6].

References

1. W. O. Alltop, An infinite class of 5-designs, *J. Comb. Theory, Ser. A* Vol. 12 (1972) pp. 390–395.
2. M. Dehon, An existence theorem for some simple t-designs, *Discrete Math.* Vol. 90 (1991) pp. 137–142.
3. S. S. Magliveras and T. E. Plambeck, New infinite families of simple 5-designs, *J. Comb. Theory, Ser. A* Vol. 44 (1987) pp. 1–5.
4. Tran van Trung, On the existence of an infinite family of simple 5-designs, *Math. Z.* Vol. 187 (1984) pp. 285–287.
5. Tran van Trung, On the construction of t-designs and the existence of some new infinite families of simple 5-designs, *Arch. Math.* Vol. 47 (1986) pp. 187–192.
6. Tran van Trung, Some existence theorems for t-designs, *Discrete Math.* Vol. 128 (1994) pp. 337–348.

Designs, Codes and Cryptography, 9, 115–120 (1996)
© 1996 Kluwer Academic Publishers, Boston. Manufactured in The Netherlands.

A New Approach to the Main Conjecture on Algebraic-Geometric MDS Codes

JUDY L. WALKER*
Department of Mathematics, University of Illinois, Urbana, IL 61801

Communicated by: V. Tonchev

Received October 13, 1994; Accepted May 5, 1995

Abstract. The Main Conjecture on MDS Codes states that for every linear $[n, k]$ MDS code over $\mathbb{F}_q$, if $1 < k < q$, then $n \leq q + 1$, except when q is even and $k = 3$ or $k = q - 1$, in which cases $n \leq q + 2$. Recently, there has been an attempt to prove the conjecture in the case of algebraic-geometric codes. The method until now has been to reduce the conjecture to a statement about the arithmetic of the jacobian of the curve, and the conjecture has been successfully proven in this way for elliptic and hyperelliptic curves.

We present a new approach to the problem, which depends on the geometry of the curve after an appropriate embedding. Using algebraic-geometric methods, we then prove the conjecture through this approach in the case of elliptic curves. In the process, we prove a new result about the maximum number of points in an arc which lies on an elliptic curve.

Keywords: Algebraic-geometric codes, MDS codes, elliptic curves

1. Introduction

Let X be an absolutely irreducible, smooth curve of genus g defined over the finite field $\mathbb{F}_q$. Let $\mathcal{P} = \{P_1, \ldots P_n\} \subset X(\mathbb{F}_q)$ be a set of n pairwise distinct $\mathbb{F}_q$-rational points on X, and let G be a divisor on X such that $\operatorname{supp} G \cap \mathcal{P} = \emptyset$ and $2g-2 < \deg G < n$. The vector space of rational functions on X associated to G is $L(G) = \{f \in \mathbb{F}_q(X) | (f) + G \geq 0\} \cup \{0\}$, where (f) is the divisor of the function f. The dimension of the divisor G is defined to be the dimension of this vector space, and is given by the Riemann-Roch theorem. See [1], [6] or [12]. The algebraic-geometric code $C = C(X, \mathcal{P}, G)$ is the image of the map

$$\begin{aligned} \alpha\colon L(G) &\to \mathbb{F}_q^n \\ f &\mapsto (f(P_1), \ldots, f(P_n)). \end{aligned}$$

C is a linear $[n, k, d]$ code with $k = \dim G = \deg G + 1 - g$ and $d \geq n - \deg G$.

The Singleton bound ([9], [12], [13]) states that the parameters of any linear $[n, k, d]$ code satisfy $d \leq n - k + 1$. In the case of equality, the code is called *maximum distance separable*, or *MDS*. A classical problem in coding theory is to find the largest possible length n of an MDS code over $\mathbb{F}_q$. To be more precise, the following conjecture can be found in [9], [13].

* Current address: Department of Mathematics and Statistics, University of Nebraska, Lincoln, NE 68588.

CONJECTURE (MAIN CONJECTURE ON MDS CODES) *For every linear $[n, k]$ MDS code over $\mathbb{F}_q$, if $1 < k < q$, then $n \leq q + 1$ except when q is even and $k = 3$ or $k = q - 1$, in which cases $n \leq q + 2$.*

The conjecture has been proven in general whenever $k \leq 5$ or $q \leq 19$. The proof for $q \leq 11$ can be found in [7], which also has a thorough exposition on arcs in projective spaces. The proofs for $13 \leq q \leq 19$ are in the recent papers [2] and [4]. In addition, the conjecture has been proven in the cases of algebraic-geometric codes arising from elliptic and, for q sufficiently large, hyperelliptic curves ([3], [5], [8], [10], [11]). These proofs depend mainly on the arithmetic of the curve, or more accurately, of its Jacobian. We will give a new proof of the conjecture in the case of elliptic curves by proving a statement about the geometry of the curve after a certain embedding.

For general references on algebraic geometry, see [1] and [6]. For algebraic-geometric codes, see [12] and [13].

2. Translation into Geometric Terms

PROPOSITION *The following are equivalent.*

(1) *For any curve $Y \subset \mathbb{P}^n_{\mathbb{F}_q}$ of degree d and genus $g \geq 1$ with $2 < n = d - g < q - 1$ and $2g + 1 \leq d < q + 2$, any set $\mathcal{Q} \subset Y(\mathbb{F}_q)$ of size $\#\mathcal{Q} = q + 2$ cannot be in general position. That is, some $n + 1$ points of $\mathcal{Q}$ must lie in a hyperplane of $\mathbb{P}^n_{\mathbb{F}_q}$.*

(2) *For any curve X of genus $g \geq 1$, there is no $[q + 2, k]$ MDS code arising from X for $3 < k < q$, except possibly for $k = g$ and $k = g + 1$.*

Proof. Let X be a curve of genus g over $\mathbb{F}_q$ and $\mathcal{P} \subset X(\mathbb{F}_q)$ with $\#\mathcal{P} = n = q + 2$. Let G be a divisor on X with $\operatorname{supp} G \cap \mathcal{P} = \emptyset$ and $2g - 2 < \deg G < q + 2$. Then $\dim G = k = \deg G + 1 - g$ by the Riemann-Roch theorem. Set $C = C(X, \mathcal{P}, G)$. Then C is not MDS if and only if the minimum distance of C is at most $n - k$ by definition. This is true if and only if there is some non-zero codeword of C having at least k coordinates equal to 0. But this is equivalent to the condition that there is some non-zero $f \in L(G)$ with $f(P) = 0$ for at least k distinct $P \in \mathcal{P}$.

(1) $\Rightarrow$ (2) With $X, \mathcal{P}, G$, and C as above, assume that $\deg G \geq 2g + 1$. This excludes the possibilities that $\deg G = 2g$ or $\deg G = 2g - 1$, which correspond to $k = \dim C = g + 1, g$ respectively. Then the divisor G is very ample and hence defines a closed embedding $\phi\colon X \to \mathbb{P}^{k-1}$ as follows ([6]): Let $\mathcal{L} = \mathcal{L}(G)$ be the invertible sheaf associated to G and let $s_0, \ldots, s_{k-1}$ be a basis for $L(G) = \Gamma(X, \mathcal{L})$, the vector space of global sections of $\mathcal{L}$. To define $\phi(P)$ for $P \in X$, let U_P be a neighborhood of P on which $\mathcal{L}(U_P) \cong \mathcal{O}_X(U_P)$, where $\mathcal{O}_X$ is the structure sheaf for X. Fix this isomorphism; say s_i corresponds to $g_{P,i}$. The isomorphism is well-defined up to a unit of $\mathcal{O}_X(U_P)$. Define $\phi(P) = (g_{P,0}(P) : \cdots : g_{P,k-1}(P))$. Letting $Y = \phi(X)$ and $\mathcal{Q} = \phi(\mathcal{P})$, we have that Y is a smooth curve of degree equal to $\deg G$ and that Y and $\mathcal{Q}$ satisfy the conditions of (1). Hence, some k points of $\mathcal{Q}$ must lie in a hyperplane. But since the points of $\mathcal{P}$ are not in the support of G, the functions s_i are regular at each point of $\mathcal{P}$. This means that the image of a point $P \in \mathcal{P}$ can

be thought of as $(s_0(P) : \cdots : s_{k-1}(P))$, and to say that k points of $\mathcal{Q}$ lie in a hyperplane is to say that for some $a_0, \ldots, a_{k-1}$, there are k distinct solutions in $\mathcal{P}$ to the equation $a_0 s_0(P) + \cdots + a_{k-1}s_{k-1}(P) = 0$. Since the function $f = a_0 s_0 + \cdots + a_{k-1}s_{k-1} \in L(G)$, this proves (2).

(2) $\Rightarrow$ (1) We prove the contrapositive of this result. So, assume that $Y \subset \mathbb{P}^n_{\mathbb{F}_q}$ is a curve of degree d and genus $g \geq 1$ with $d = n + g$, $2 < n < q - 1$, and $2g + 1 \leq d < q + 2$, and let $\mathcal{Q} \subset Y(\mathbb{F}_q)$ be a set of $q + 2$ points in general position. We will show that there is a $[q+2, n+1]$ MDS code arising from Y and $\mathcal{Q}$. Let $H.Y$ be any hyperplane section. By the approximation theorem ([12]), there is a divisor G defined over $\mathbb{F}_q$ which satisfies (1) $G \sim H.Y$ and (2) $\operatorname{supp} G \cap \mathcal{Q} = \emptyset$. Then G is a divisor of degree d. Consider the code $C = C(Y, \mathcal{Q}, G)$ of length $q + 2$ and dimension $k = d + 1 - g = n + 1$. We know that C is not MDS if and only if there is some $f \in L(G)$ with $f(Q) = 0$ for at least k distinct $Q \in \mathcal{Q}$. But if such an f exists, we must have $(f) + G = H'.Y$ for some hyperplane H' since the set of hyperplane sections is a complete linear system on Y. But then H' contains at least $k = n + 1$ points of $\mathcal{Q}$, which is impossible since $\mathcal{Q}$ is in general position.

■

From the proof, we see that proving that there is no $[q + 2, k]$ MDS code arising from a given curve X is equivalent to proving that when X is embedded in $\mathbb{P}^{k-1}$ as a curve of degree $k+g-1$, no $q+2$ $\mathbb{F}_q$-rational points of X can be in general position. It can easily be shown that if there is an $[n, k]$ MDS code arising from X with $n > q+2$, then there is a $[q+2, k]$ MDS code arising from X, see ([10]). Since the dual of an algebraic-geometric MDS code arising from X and $\mathcal{P}$ is an algebraic-geometric MDS code arising from X and $\mathcal{P}$, in order to prove that the main conjecture on MDS codes holds for codes coming from a curve X, it is enough to prove that condition (1) holds for curves $Y \cong X$ and $n = k - 1 \leq \lfloor \frac{q}{2} \rfloor$.

3. A Statement About *k*-arcs

We first prove a stronger result.

THEOREM 1 *Fix $\delta > \frac{2}{3}$. Then there exists $q_0 = q_0(\delta)$ such that whenever $q \geq q_0$, $m \geq \delta q$, $Y \subset \mathbb{P}^n$ is a curve of genus one and degree $n + 1$ defined over $\mathbb{F}_q$ with $n \leq \lfloor \frac{m}{2} \rfloor - 1$, and $\mathcal{Q} \subset Y(\mathbb{F}_q)$ is a set of m rational points on Y, then some $n + 1$ points of $\mathcal{Q}$ must lie in a hyperplane of $\mathbb{P}^n$.*

Remark. In the language of finite geometries, this result states the following:

Let $Y \subset PG(n, q)$ be an algebraic curve of genus 1 and degree $n + 1$. Then as $q \to \infty$, the largest size of a k-arc contained in Y approaches $\frac{2}{3}q$.

In order to prove this theorem, we will need the following technical fact. The proof follows an idea used by Munuera ([10]).

LEMMA *Fix $\delta > \frac{1}{2}$. Then there exists $q_0 = q_0(\delta)$ such that whenever $q \geq q_0$, X is a curve of genus one defined over $\mathbb{F}_q$, and $\mathcal{P} \subset X(\mathbb{F}_q)$ is a set of $m \geq \delta q$ rational points on X,* *there is a divisor class* $[D]$ *consisting of divisors of degree 2 that has at least* $\lfloor \frac{m}{4} \rfloor + 2$ *representatives of the form* $P + P'$*, where* P *and* P' *are distinct points in* $\mathcal{P}$.

Proof of Lemma: Since the genus of X is one, the number of divisor classes of degree 2 of X is the number of points of degree one on X, N_1. Suppose the lemma is not true. Then each divisor class $[D]$ of degree 2 has at most $\lfloor \frac{m}{4} \rfloor + 1$ representatives of the desired form. Since the number of divisors of the correct form is $\binom{m}{2}$, we have

$$\binom{m}{2} \leq N_1 \left(\left\lfloor \frac{m}{4} \right\rfloor + 1 \right).$$

Using Serre's improvement on the Hasse-Weil bound, we have

$$\binom{m}{2} \leq (q + 1 + \lfloor 2\sqrt{q} \rfloor) \left(\left\lfloor \frac{m}{4} \right\rfloor + 1 \right).$$

We rewrite this inequality as

$$\frac{m^2}{2} - q \left\lfloor \frac{m}{4} \right\rfloor \leq \lfloor 2\sqrt{q} \rfloor \left\lfloor \frac{m}{4} \right\rfloor + \frac{m}{2} + \left\lfloor \frac{m}{4} \right\rfloor + q + \lfloor 2\sqrt{q} \rfloor + 1.$$

The left-hand side of this inequality is quadratic in q with positive coefficient as long as $m \geq \delta q$ with $\delta > \frac{1}{2}$, and the largest power of q on the right-hand side is $q^{3/2}$, so the inequality cannot hold for q sufficiently large. ∎

Note that the supports of these divisors in $[D]$ are all disjoint, for if $P + P' \sim P + P''$, then $P' \sim P''$, which is impossible unless $P' = P''$.

Proof of Theorem 1: Let $r = n - 1$ if n is odd, and $n - 2$ if n is even. Let $Q_1, \ldots, Q_{n-1}$ be distinct points in $\mathcal{Q}$ chosen so that $Q_1 + Q_2 \sim Q_3 + Q_4 \sim \cdots \sim Q_{r-1} + Q_r \in [D]$, where $[D]$ satisfies the property described in the lemma. Since $\frac{r}{2} \leq \lfloor \frac{\lfloor \frac{m}{2} \rfloor - 2}{2} \rfloor = \lfloor \frac{m}{4} \rfloor - 1$, the lemma says that there are at least three more divisors of the form $P + P' \in [D]$ with P, P' distinct points of $\mathcal{Q}$. Note that even in the case of n even and Q_{n-1} in the support of one of these three divisors, we still have two more divisors of the desired form in $[D]$ with support disjoint from $\{Q_1, \ldots, Q_{n-1}\}$. Let $\mathcal{Q}' = \mathcal{Q} \setminus \{Q_1, \ldots, Q_{n-1}\}$. Note that for each $Q \in \mathcal{Q}'$, we may assume that there is a unique hyperplane H_Q containing $Q, Q_1, \ldots, Q_{n-1}$, since otherwise $Q_1, \ldots Q_{n-1}, Q$ span at most a copy of $\mathbb{P}^{n-2}$ and we can add any other point of $\mathcal{Q}$ to get $n + 1$ points of $\mathcal{Q}$ contained in a hyperplane. Further, since Y is a curve of degree $n + 1$, the hyperplane section $H_Q.Y$ has degree $n + 1$. Hence, the divisor $H_Q.Y - Q - Q_1 - \cdots - Q_{n-1}$ is an effective divisor of degree 1, i.e. a point of $Y(\mathbb{F}_q)$. Let $f\colon \mathcal{Q}' \to Y(\mathbb{F}_q)$ be the map of sets which takes Q to $H_Q.Y - Q - Q_1 - \cdots - Q_{n-1}$. We will show that for some Q, $f(Q) \in \mathcal{Q}' \setminus \{Q\}$. This would mean that $H_Q.Y$ would be a sum of $n + 1$ distinct points of $\mathcal{Q}$, and hence that the hyperplane H_Q would contain $n + 1$ points of $\mathcal{Q}$.

First, suppose that $f(Q) = Q_i$ for some i with $1 \leq i \leq r$ and some $Q \in \mathcal{Q}'$. For simplicity of notation, assume $i = 1$. Then there is some divisor $P + P'$ with $P, P' \in$

$\mathcal{Q}' \setminus \{Q\}$ and $P + P' \in [D]$. In particular $P + P' \sim Q_1 + Q_2$, and so $H_Q.Y = 2Q_1 + Q_2 + \cdots + Q_{n-1} + Q \sim Q_1 + Q_3 + \cdots + Q_{n-1} + Q + P + P'$. But then this last divisor is a hyperplane section since the set of hyperplane sections form a complete linear system on Y, and its support consists of $n+1$ distinct points from $\mathcal{Q}$. So in this case, we are done. Therefore, from now on we assume that $f(\mathcal{Q}') \subset Y(\mathbb{F}_q) \setminus \{Q_1, \ldots, Q_r\}$.

Next, we must count the number of $Q \in \mathcal{Q}'$ such that $f(Q) = Q$. To do this, we will define a finite separable $\bar{\mathbb{F}}_q$-morphism $\pi\colon \bar{Y} \to \mathbb{P}^1$ and use Hurwitz's Theorem; see, for example, [1] or [6]. Here $\bar{\mathbb{F}}_q$ is the algebraic closure of $\mathbb{F}_q$ and $\bar{Y}$ is the curve obtained from Y by extending scalars to $\bar{\mathbb{F}}_q$. Fix some copy of $\mathbb{P}^1 \subset \mathbb{P}^n$ which does not contain $Q_1, \ldots, Q_{n-1}$. Let V be the copy of $\mathbb{P}^{n-2}$ spanned by $Q_1, \ldots, Q_{n-1}$. We may assume that V does not contain any other points of $\mathcal{Q}$, since otherwise we would be done. Let $S = \bar{Y} \cap V$. Then S is a finite set of points of $\bar{Y}$, and $S \cap \mathcal{Q}' = \emptyset$. Let $X = \bar{Y} \setminus S$. Then X is a nonempty, open, dense subest of $\bar{Y}$, and a quasi-projective curve in its own right. We define $\pi_0\colon X \to \mathbb{P}^1$ to be projection from V. Then π_0 is a morphism which is given topologically as follows: For a point $P \in X$, $\pi_0(P)$ is the unique point where the unique hyperplane determined by V and P meets $\mathbb{P}^1$. Note that two points of X get sent to the same point of $\mathbb{P}^1$ if and only if they, along with V, are contained in a hyperplane. By repeated application of [6, I.6.8], the map π_0 extends uniquely to our desired morphism $\pi\colon \bar{Y} \to \mathbb{P}^1$. Since the degree of Y is $n+1$, at most two points of X can go to the same point of $\mathbb{P}^1$. Further, if for some $P \in X$, we have $\pi_0^{-1}(\pi_0(P))$ consisting of only the one point P, then the hyperplane spanned by P and V must be tangent to Y at P. Since $\pi|_X = \pi_0$, π is finite of degree 2. Further, if π is not separable, then it is purely inseparable, which is impossible by [6, IV.2.5]. Hence Hurwitz's theorem applies to show that the number of ramified points of π is at most 4. Now, by the definition of f, we have $f(Q) = Q$ if and only if the hyperplane spanned by $Q_1, \ldots, Q_{n-1}, Q$ is tangent to Y at Q. This is equivalent to the statement that $f(Q) = Q$ for $Q \in \mathcal{Q}'$ if and only if π_0, and hence π, ramifies at Q. Therefore, at most 4 points $Q \in \mathcal{Q}'$ can have the property $f(Q) = Q$.

Note that f is one-to-one. In fact, if $f(P) = f(P') = O$, say, then we have $Q_1 + \cdots + Q_{n-1} + P + O = H_P.Y \sim H_{P'}.Y = Q_1 + \cdots + Q_{n-1} + P' + O$. Hence $P \sim P'$, which is impossible unless $P = P'$. So the number of points in the image of f is exactly $\#\mathcal{Q}' = m - (n-1) = m - n + 1$.

Finally, if no point $Q \in \mathcal{Q}'$ has the property that $f(Q) \in \mathcal{Q}' \setminus \{Q\}$, then

$$\#\mathcal{Q}' + \#f(\mathcal{Q}') - 4 \le \#(\mathcal{Q}' \cup f(\mathcal{Q}')) \le N_1 - r$$

by inclusion-exclusion. Using Serre's improvement on the Hasse-Weil bound, we have

$$2(m - n + 1) - 4 \le q + 1 + \lfloor 2\sqrt{q} \rfloor - r,$$

which is the same as

$$2m - 2n - q + r - 3 \le 2\sqrt{q}.$$

Using the facts that $n \le \lfloor \frac{m}{2} \rfloor \le \frac{m}{2}$ and $r \ge n - 2$, we rewrite this inequality as

$$\frac{3m}{2} - q - 5 \le \lfloor 2\sqrt{q} \rfloor.$$

The left-hand side of this last inequality is linear in q as long as $m \geq \delta q$ with $\delta > \frac{2}{3}$, and so we have a contradiction for q sufficiently large. ■

4. A New Proof of the Conjecture for Genus 1

THEOREM 2 *The main conjecture on MDS codes holds for all algebraic-geometric codes coming from elliptic curves defined over* $\mathbb{F}_q$.

Proof. Since the main conjecture has been proven using the theory of finite geometries whenever $k \leq 5$ or $q \leq 19$, it is enough to prove the conjecture for $k \geq 6$ and $q > 19$. Further, since $2g + 1 = 3 < 5$ for an elliptic curve, it is enought to show that (1) holds whenever Y has genus 1 and $5 \leq n \leq \lfloor \frac{q}{2} \rfloor$. Hence, it is enough to show that the assumptions of the Lemma and Theorem 1 hold when we set $m = q + 2$ and $q > 19$. This is easily checked, and the theorem is proven. ■

Acknowledgement

The author thanks the second referee for pointing out the references [2] and [4].

References

1. S. S. Abhyankar, *Algebraic Geometry for Scientists and Engineers*, American Mathematical Society, Providence, Rhode Island (1990).
2. A. Ali, J. W. P. Hirschfeld and H. Kaneta, On the size of arcs in projective spaces, *IEEE Trans. Inform. Theory*, to appear.
3. M. A. de Boer, MDS codes from hyperelliptic curves, preprint.
4. J. M. Chao and H. Kaneta, Rational arcs in $PG(r, q)$ for $11 \leq q \leq 19$, preprint.
5. H. Chen, On the main conjecture of hyperelliptic MDS codes, preprint.
6. R. Hartshorne, *Algebraic Geometry*, Springer-Verlag, New York-Heidelberg-Berlin (1977).
7. J. W. P. Hirschfeld and J. A. Thas, *General Galois Geometries*, Oxford University Press, Oxford (1991).
8. G. L. Katsman and M. A. Tsfasman, Spectra of algebraic-geometric codes, *Probl. Peredachi Inform.*, Vol. 23 (1987) pp. 19–34.
9. F. J. MacWilliams and N. J. A. Sloane, *The Theory of Error-Correcting Codes*, North-Holland, Amsterdam (1977).
10. C. Munuera, On the main conjecture on geometric MDS codes, *IEEE Trans. Inform. Theory*, Vol. 38 (1992) pp. 1573–1577.
11. L. Søndergard, The non-existence of long MDS codes from elliptic curves, preprint.
12. H. Stichtenoth, *Algebraic Function Fields and Codes*, Springer-Verlag, Berlin-Heidelberg-New York (1991).
13. M. A. Tsfasman and S. G. Vlăduţ, *Algebraic-Geometric Codes*, Kluwer Academic Publishers, Dordrecht-Boston-London (1991).

GPSR Compliance

The European Union's (EU) General Product Safety Regulation (GPSR) is a set of rules that requires consumer products to be safe and our obligations to ensure this.

If you have any concerns about our products, you can contact us on ProductSafety@springernature.com

In case Publisher is established outside the EU, the EU authorized representative is:

Springer Nature Customer Service Center GmbH
Europaplatz 3
69115 Heidelberg, Germany

Batch number: 10755292

Printed by Printforce, the Netherlands